T0227403

Multiphase Biomedical Materials

Titles of Related Interest

Books

New Polymeric Materials: Reactive Processing and Physical Properties
Editors: E. Martuscelli and C. Marchetta

History of Polymeric Composites
Editors: R. B. Seymour and R. D. Deanin

Advanced Composite Materials and Structures
Editors: G. C. Sih and S. E. Hsu

New Concepts in Polymer Science
Series Editor: C. R. H. I. de Jonge

Journals

Journal of Biomaterials Science—Polymer Edition
Editors: T. Tsuruta, S. L. Cooper and C. H. Bamford

Artificial Organs Today
Editor: T. Agishi

Frontiers of Medical and Biological Engineering
Editor: M. Saito

Journal of Adhesion Science and Technology
Editors: K. L. Mittal and W. J. van Ooij

New Polymeric Materials
Editor: F. Karasz

Multiphase Biomedical Materials

Edited by

T. TSURUTA
Science University of Tokyo, Japan

and

A. NAKAJIMA
Osaka Institute of Technology, Japan

CRC Press
Taylor & Francis Group
Boca Raton   London   New York

CRC Press is an imprint of the
Taylor & Francis Group, an **informa** business

© 1989 VSP BV

First published in 1989
ISBN 90-6764-109-X

CIP-DATA KONINKLIJKE BIBLIOTHEEK, DEN HAAG

Multiphase biomedical materials/ed. by T. Tsuruta
and A. Nakajima.—Zeist: VSP.—Ill.
ISBN 90-6764-109-X bound
SISO 601.9 UDC 614-74/-77
Subject heading: biomedical materials.

Typeset in Great Britain by Blackpool Typesetting Services Ltd., Blackpool

Contents

Subject index

Preface

Since microheterogeneity (Y. Imai and E. Masuhara) or micro-architecture (D. J. Lyman) of the surface of materials was seen as playing an important role in determining the mode of cell material interactions, there were quite a few reports presented in the 1970s and 1980s suggesting correlations of multiphase structures of the surface of materials with their antithrombogenicity or biocompatibility.

A Special Research Project 'Design of Multiphase Biomedical Materials', which was supported by the Japanese Ministry of Education, Science and Culture, was promoted between 1982 and 1986 in Japan under interdisciplinary cooperation among researchers in materials science and the biological and medical sciences. The objective of this research project was to elucidate various aspects of biomedical behavior of multiphase systems at the interface with living bodies at the molecular, cellular and tissue levels. Multiphase materials studied in the research project cover polymers having microphase-separated structures, hydrogels, immobilized enzymes (or cells), ceramics, and metallic materials.

The research project was carried out by five subgroups: (I) multiphase biomedical materials with microdomain structures; (II) multiphase biomedical materials containing liquid components; (III) hybrid-type multiphase biomedical materials with biological components; (IV) inorganic and metallic multiphase biomedical materials; and (V) methods for analysis and evaluation of multiphase biomedical materials.

Thanks to the great efforts made by members of the research project, a marked progress of basic and applied studies has been achieved in the field of biomaterials science and technology. Especially, new concepts obtained in the course of the research project are now exploring new frontiers and innovations in the expanding world of biomaterials.

This book was edited to provoke a better understanding about various aspects of cell–material interactions in whch the multiphase systems play a crucial role. The contents are organized by selecting eleven high-light researches from the five subgroups as follows:

Chapters	1	2	3	4	5	6	7	8	9	10	11
Subgroups	I, V	I	II	V	III	III	IV	IV	III	III	II

Each writer of the ten chapters describes the results of his own research, featuring the objectives and characteristics of the subgroup project.

The Editors express their sincere acknowledgment to the Japanese Ministry of Education, Science and Culture, for its support to the Special Research Project, which brought about fruitful results and motivated us to edit this book in order to provoke international recognition of the importance of multiphase biomedical materials.

T. Tsuruta and A. Nakajima
Editors

Multiphase Biomedical Materials, pp. 1–19 (1989)
T. Tsuruta and A. Nakajima (Eds)
© 1989 VSP.

Chapter 1

Controlled interactions of cells with multiphase-structured surfaces of block and graft copolymers

KAZUNORI KATAOKA,*[1] TERUO OKANO,[1] YASUHISA SAKURAI,[1] ATSUSHI MARUYAMA[†2] and TEIJI TSURUTA[2]

[1]*Institute of Biomedical Engineering, Tokyo Women's Medical College, 8-1 Kawada, Shinjuku, Tokyo 162, Japan*
[2]*Department of Industrial Chemistry, Science University of Tokyo, Kagurazaka, Shinjuku, Tokyo 162, Japan*

Summary—Detailed studies on cell-material interaction greatly facilitate the molecular design of materials having a regulative effect on cellular functions. These type of materials will have novel applications in the diverse field of biomedical sciences and technologies. This paper reviews the promising features of microdomain-structured polymers as materials capable of regulating several cellular functions. Particular emphasis is placed on the successful utilization of these microdomain-structured polymers as innovative biofunctionality materials such as non-thrombogenic materials and cell-specific adsorbents.

1. INTRODUCTION

Recently, cellular behavior at material interfaces has received a great deal of interest from many researchers in the field of biomaterials science. A principal motivation of this interest is due to the consensus that regulation of cellular behavior at the surface of biomaterials may open a new pathway to develop advanced biomaterials for innovations in the field of clinical medicine as well as in the various biotechnologies including gene, protein and cell engineering.

Indeed, there have been many great efforts to obtain an insight into the physical and chemical parameters of materials which crucially affect aspects of cellular behavior such as adhesivity, shape change, functional alteration and proliferation. These efforts have been more or less based on the classical colloidal theory, in which cells are regarded as homogeneous particles. Such parameters as wettability and surface potential, reflecting the averaged physical

---

* Present address: Department of Materials Science and Research Institute for Biosciences, Science University of Tokyo, Noda, Chiba 278, Japan
† Present address: Department of Chemistry, Sophia University, Kioi-cho, Chiyoda-ku, Tokyo 102, Japan.

character of the surface, have often been applied to understanding cellular behavior at interfaces, although only limited success has been achieved in terms of establishing a generalized theory to regulate cellular behavior at interfaces.

As is well known, the cellular plasma membrane has a highly heterogeneous or mosaic structure composed of different types of molecules, the main constituents of which are glycoproteins and phospholipids [1]. An important characteristic of the plasma membrane is that most of its components are capable of lateral mobility in the membrane plane [2]. As a result of this mobile nature, membrane glycoprotein molecules can form clusters, as confirmed by transmission electron microscopic (TEM) observation of freeze-fractured membranes. In this sense, one may refer to cells as molecular assemblages with flexible membranes rather than as rigid particles.

Cluster formation of glycoprotein molecules at the membrane surface is considered to be strongly coupled with the dominant cellular functions. One may see a typical example in the interaction of lymphocytes with multivalent proteinaceous ligands including lectins and antibodies. Lymphocytes undergo activation due to the binding of these ligands to the glycoprotein molecules on the plasma membrane surfaces. Ligand-induced clustering of glycoprotein molecules, known as 'patching' and 'capping', is believed to trigger the transmembrane response that results in the change in the intracellular metabolism [3].

Considering the dynamic and flexible nature of cellular plasma membrane, the cell–material interaction must take place at multiple points (multipoint interaction). It seems likely that multipoint interaction, or plane-to-plane interaction, may lead to the lateral rearrangement of membrane constituents, including glycoprotein molecules, at the membrane plane contacting with the materials surface. The mode of rearrangement would be strongly affected by the properties characteristic to the contacting plane of the materials surface. It does not seem unreasonable to suppose that microheterogeneity of the material surface may take a crucial role in molecular rearrangement or cluster formation of the membrane constituents, and thus, may dominate a cellular response at the material interface [4, 5].

This concept of 'capping-control' is supported through our systematic study on the interaction of cells with the microdomain-structured surfaces of a series of block and graft copolymers. This paper reviews a unique nature of those microdomain- or mosaic-structured surfaces in terms of their response to several kinds of mammalian cells including platelets and lymphocytes. Further, particular emphasis is placed on the successful utilization of these microdomain-structured materials as high-value-added specialty biomaterials including antithrombogenic materials and specific adsorbents for a particular lymphocyte subpopulations (artificial cellular-receptor).

2. MOLECULAR DESIGN OF NON-THROMBOGENIC POLYMERS

2.1. Non-thrombogenic polymers
Polymeric materials have contributed significantly to the development and advancement of devices and systems in biomedical fields. Presently, development

of ideal non-thrombogenic polymers, an important problem in biomaterials science, is eagerly desired in the fields of cardiovascular prostheses, artificial heart research and other blood-contacting devices and systems.

When a polymer surface is in contact with blood, plasma proteins immediately adsorb on the polymer surface and soon thereafter, platelets adhere on the proteinated surface. The adhered platelets extend pseudopods and release ATP, thromboxane, granules, etc., and the platelets then aggregate with a fibrin net to form a white clot. It is believed that these initial processes govern the nature of the subsequent thrombus formation and influence the thrombogenecity of the polymer over the long term. Discussions of antithrombogenic polymers from the stand point of interactions between polymers and proteins or platelets [6] should prove valuable in the development of such materials.

2.2. Microdomain concept

Block copolymers having hydrophilic–hydrophobic microdomain structures are found to exhibit non-thrombogenic activity, both *in vivo* and *ex vivo*, due to a marked suppression of activation of adhering platelets [7, 8]. From this perspective, the effect of morphology [8], size [9] and chemical structure [8–11] on the interaction of polymer surfaces with blood elements such as protein, platelets, etc., have been studied.

ABA type block copolymers constructed by 2-hydroxyethyl methacrylate (HEMA) (*A*) and styrene (St) (*B*) or dimethylsiloxane (DMS) (*B*) have been synthesized by a coupling reaction as described in previous papers [10–12]. The number and the morphology of adhered platelets was investigated by a column method [6]. In addition, the internal surfaces of a polyester–polyurethane tube (1.5 mm i.d. × 20 cm in length) were coated with copolymers and applied *ex vivo* as an arterio-venous (A-V) shunt, i.e. connecting the carotid artery to the jugular vein via the tube. *Ex vivo* non-thrombogenecity of the copolymers was evaluated by monitoring the blood flow rate to determine the occlusion by thrombosis [7]. *Ex vivo* non-thrombogenecity of polymers could thus be evaluated by the rate of occlusion by thrombosis. The occlusion time was determined as the time at which the flow rate decreased to zero. The mean occlusion times of the polymers tested were shown in Table 1. The block copolymers showed excellent non-thrombogenic character; the occlusion time was 20 days for the HEMA-St block copolymer containing 0.61 mol fraction of HEMA and

Table 1.
Platelet adhesion and *in vivo* antithrombogenicity

Polymer	*In vivo* antithrombogenicity occlusion time (days)[a]	Platelet adhesion (%)	Platelet shape change
HEMA-St	20 ± 2.0	11.8	round
HEMA-DMS	12 ± 1.5	63.5	round
PHEMA	3 ± 1.5	33.5	deformed
PSt	2 ± 1.0	38.6	deformed
PDMS	3 ± 1.0	40.0	deformed

[a] Mean ± SD (*n* = 7).

12 days for the HEMA–DMS block copolymer containing 0.67 mol fraction of HEMA. The HEMA–St block copolymer containing 0.61 mol fraction of HEMA exhibited a hydrophilic–hydrophobic microdomain structure where the morphology was highly ordered, alternate lamella, as shown in a previous paper [9–12]. In the *in vitro* examinations, the surface of the HEMA–St block copolymer was found to have an ability to suppress platelet adhesion and plate-let activation, i.e. shape change and release [7–11]. This suppressing effect does not appear on homopolymer surfaces. The HEMA–DMS block copolymer containing 0.67 mol fraction of HEMA also formed the hydrophilic–hydrophobic microdomain structure of modified lamella as shown previously [9]. With this HEMA–DMS block copolymer, the platelet adhesion was not suppressed; however, the activation of adhering platelets was suppressed remarkably.

The rapid adsorption of plasma proteins is the initial event in a complex series of reactions that occurs when polymeric materials contact blood. This adsorbed protein layer is considered either to accelerate or to retard the reaction of thrombus formation. Therefore, characteristics of the adsorbed protein layer formed on hydrophilic–hydrophobic microdomain surfaces must be an impor-tant clue to elucidate the role of the microdomain structure in suppressing the activation of adhering platelets.

In a mono-protein solution system, distribution of adsorbed protein onto the hydrophilic–hydrophobic microdomain surface of HEMA–St were examined by electron microscopy, using the osmium tetroxide fixation technique. The results showed that specific adsorption of proteins on the microdomains had occurred. Albumin selectively adsorbed on hydrophilic microdomains, while γ-globulin and fibrinogen selectively adsorbed on hydrophobic microdomains [13].

In a binary protein solution system, the distribution of adsorbed albumin and γ-globulin formed organized structures corresponding to the microdomain surface structures of the block copolymer. That is, albumin selectively adsorbed to hydrophilic microdomains [14].

The organized protein layer formed on the microdomain-structured surface is considered to have the ability to suppress the activation of adhering platelets due to regulation of the distribution of binding sites between platelets and the block copolymer surface, as shown in Fig. 1.

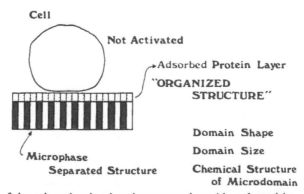

Figure 1. Role of the polymeric-microdomain structure in antithrombogenicity.

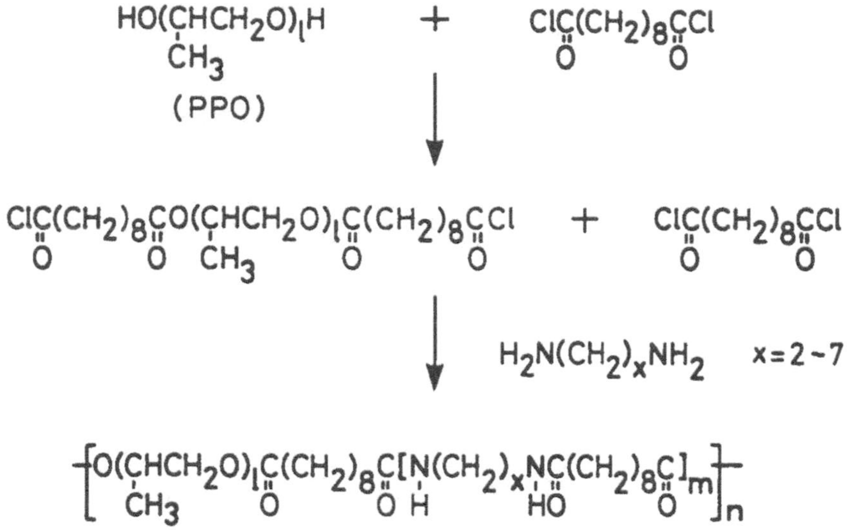

$$HO(CHCH_2O)_lH \qquad + \qquad ClC(CH_2)_8CCl$$
$$\quad\ CH_3 \qquad\qquad\qquad\quad\ \overset{\|}{O} \qquad\quad \overset{\|}{O}$$
$$(PPO)$$

$$ClC(CH_2)_8CO(CHCH_2O)_lC(CH_2)_8CCl \qquad + \qquad ClC(CH_2)_8CCl$$
$$\overset{\|}{O} \qquad\ \overset{\|}{O}\ \ CH_3 \quad\ \overset{\|}{O} \qquad \overset{\|}{O} \qquad\qquad\qquad \overset{\|}{O} \qquad \overset{\|}{O}$$

$$H_2N(CH_2)_xNH_2 \qquad x = 2 \sim 7$$

$$\left[O(CHCH_2O)_lC(CH_2)_8C[N(CH_2)_xNC(CH_2)_8C]_m \right]_n$$
$$\qquad CH_3 \quad\ \overset{\|}{O} \quad\ \ \overset{\|}{O}\ \overset{|}{H} \qquad\quad \overset{|}{H}\overset{|}{O}\ \ \overset{\|}{O}$$

Figure 2. Synthetic route of a PPO-segmented polyamide.

2.3. Non-thrombogenecity of multiphase polymeric materials with crystalline/amorphous microdomain structure

Apart from amorphous block copolymers including HEMA–St and HEMA–DMS, excellent antithrombogenic polymers have also been found in the category of block copolymers having crystalline or paracrystalline segments, although their microdomain-structure morphology is distinct from that of amorphous copolymers. A well-known example of this type of material is segmented poly(ether urethaneurea) (PEUU), from which pneumatic-driven artificial heart was manufactured [15].

PEUU is considered to have a microdomain structure of the so-called fringed-micelle type: clusters of hard segments are dispersed in a continuous phase of soft segments to form the microdomains. Nevertheless, the relation of the PEUU microstructure to its antithrombogenecity has not been fully understood yet. In order to clarify the role of the crystalline-amorphous type microdomain structure in antithrombogenecity, we have designed segmented polymers with a well-defined microdomain structure [16, 17]. As shown in Fig. 2, these copolymers were composed of poly(propylene oxide) (PPO) and aliphatic polyamide segments. The PPO segment is wholly amorphous, whereas the other, aliphatic polyamide (nylon), is partially crystallized to form spherulites in the matrix of the amorphous segments. This type of microstructure is characterized by X-ray analysis, from which one can determine the crystalline thickness (D_{002}) and the long period (L), a parameter representing the distance between crystallites (see Fig. 3).

Of interest, canine platelet adhesion to these copolymer surfaces was found to be closely coupled with the copolymer microstructure. As shown in Figs 4 and 5, platelet adhesion took the least time on the copolymer surfaces having

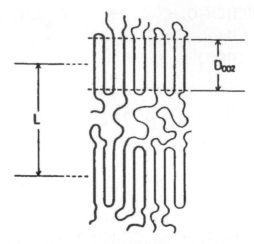

Figure 3. Crystalline/amorphous microstructure of a PPO-segmented polyamide.

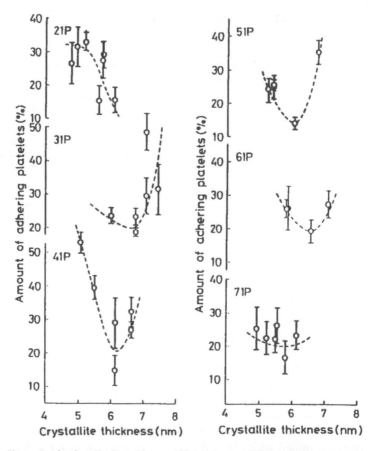

Figure 4. Change in platelet adhesion with crystallite thickness of PPO-segmented polyamides. The *disit* preceding *IP* represents the methylene chain length of the diamine unit.

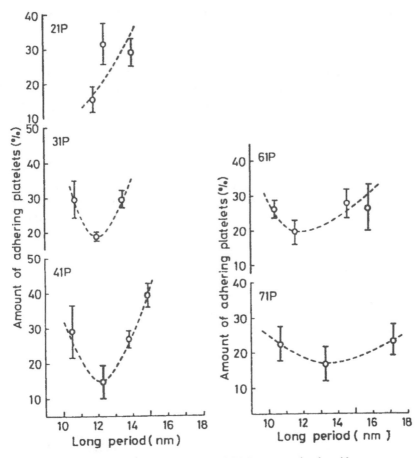

Figure 5. Change in platelet adhesion over time of PPO-segmented polyamides.

a crystallite thickness of 6.0–6.5 nm and long periods of 12–13 nm [17]. It is to be noted that these optimal ranges in a long period and a crystallite thickness are common in every copolymer with varying number of methylene units in polyamide segments examined. These results make it clear that the size and distribution of crystalline and amorphous phases in the copolymer are the most crucial factors regulating platelet adhesion on these copolymers. This type of microstructure formation is driven through the polyamide crystallization that results in the continuity of the same microstructure, and thus, the same chemical composition, from the interior to the surface of the copolymer. This isotropic distribution of crystalline and amorphous phases was confirmed by X-ray photoelectron spectroscopy (XPS) and TEM measurements [18, 19]. This isotropic feature of the polyamide microstructure lends strong support to our concepts that platelets restrain themselves from activation through their recognition of the microdomain structure of the surfaces of these segmented polyamides.

3. MOLECULAR DESIGN OF MATERIALS FOR CELL SEPARATION

3.1. Cellular adsorption chromatography

In order to develop a cellular adsorption chromatography system which separates cells upon a difference in cellular adhesivity to a substrate surface, substrates having an ability to adsorb cells without giving them undesirable side effects or damages are essential. The adsorption of cells on a foreign substrate is able to be discriminated to two different but serial processes as illustrated in Fig. 6 [5]. The first stage is the physicochemical adsorption process (passive adhesion process), where cellular adsorption is primarily determined by the physicochemical nature of both the cellular plasma membrane surface and the substrate surface. The second process is the active adhesion process. This process is closely related to cellular energy metabolism. In this process, cellular adsorption on the substrate is further enhanced through some metabolic changes, accompanied with cellular shape changes such as flattening and pseudopod formation which are triggered by cellular contact with the substrate surface (contact induced activation). Undesirable side effects on cells is also brought about from this process. Thus, the substrate used for cellular adsorption chromatography should prevent adsorbed cells proceeding to the second stage. Further, this preventive effect of the substrate enables one to discriminate cells based on differences in their physicochemical affinities, i.e. difference in cellular adhesivity in the first stage.

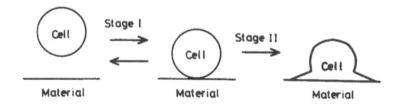

Stage I : Adsorption Process (Passive Process)

Stage II : Adhesion Process (Active or energy-requiring Process)

Figure 6. Two stages of cellular interaction with materials [5].

3.2. Cellular interaction with microdomain-structured surfaces of polyamine graft copolymers

We have been carrying out a series of studies concerning with cellular interactions with various types of synthetic polymer substrate in terms of cellular adhesion behavior [20–22]. We have found that polymer substrates having microphase separated structures retain blood cells without inducing their shape changes. As shape changes of cells are known to be closely associated with changes in function and to require cellular metabolic changes [23, 24], we considered the substrate which adsorb cells without inducing shape change did not trigger cellular activation nor functional change. Thus, to gain further

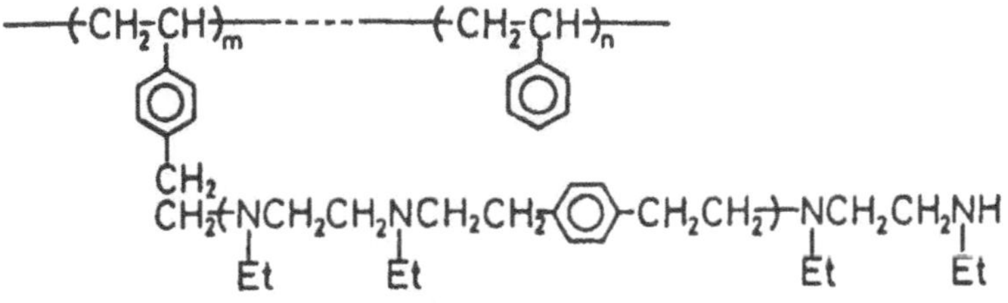

$$-\text{(CH}_2\text{CH)}_m \ ---- \ -\text{(CH}_2\text{CH)}_n-$$

SAX

$$-\text{(CH}_2\text{CH)}_m \ ---- \ -\text{(CH}_2\text{C)}_n-$$

$$\text{O=COCH}_2\text{CH}_2\text{OH}$$

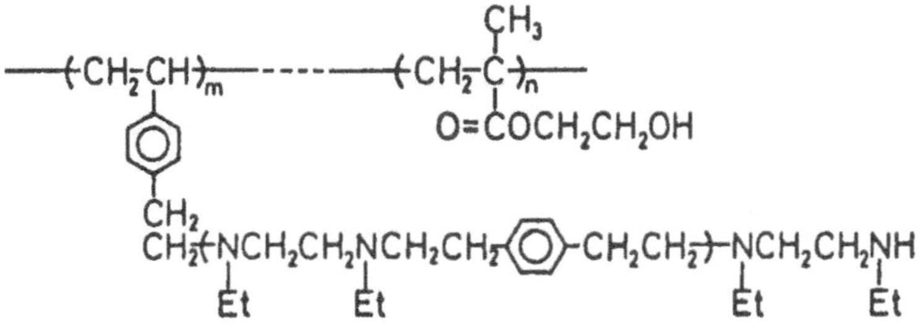

HAX

Figure 7. Structural formulas of graft copolymers. '*x*' represents the wt.% of polyamine portion in the copolymer.

information, polymer substrates having a microphase-separated structure containing microdomains of polyamine, which has a relatively strong interaction with cells, were prepared and their interactions with blood cells were evaluated. Their structural formulas are shown in Fig. 7. These graft copolymers with polyamine grafts have been prepared by utilizing the macromonomer concept [25]. We can get various types of polyamine graft copolymers by radical copolymerization of polyamine macromonomers with various comonomers [26, 27]. Thus, graft copolymers with a backbone polymer of the desired character are readily prepared. An electron microscopic observation of thin films of the polymers clearly shows that the copolymers form a sea–island type microphase-separated structure [27, 28]. The continuous 'sea-like' domains consist of the backbone polymer and 'island-like' domains consist of polyamine grafts of the copolymer. The microdomains are in the range of from 10^{-2} to 10^{-1} μm in size; their size dimension is considerably smaller than that of blood cells. Namely, when cells are in contact with the copolymer surface, they should come into contact with plural microdomains of the polymer at once, and multipoint interactions between the cellular surface and polymer surface would take place.

Evaluation of the interaction between the polymer substrates and blood cell such as platelets and lymphocytes was carried out by employing the microsphere column method: a cell suspension was passed through a column packed with polymer coated glass beads. The extent of cellular interaction with a substrate

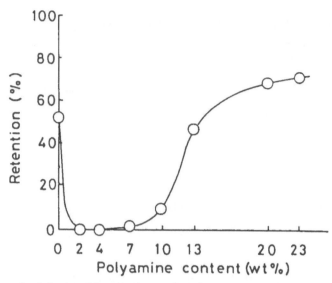

Figure 8. Retention behavior of lymphocytes on the HA copolymer.

is measured by means of values of percentage retention and of percentage effusion. Both values can be calculated from the concentration of the cell suspension before and after passage through the column. The value of the percentage retention of rat mesenteric lymphocytes obtained for poly(2-hydroxyethyl methacrylate)–polyamine graft copolymers (HAx copolymers, x indicates wt.% of polyamine graft in the copolymer) with various polyamine contents are summarized in Fig. 8 [29]. It is to be noted that lymphocyte retention on the copolymers is significantly depressed on copolymers with a polyamine content of 4–7 wt.%. Further introduction of polyamine graft over 7 wt.% induced an increase in lymphocyte retention on the copolymer. Morphological observation on lymphocytes adsorbed on the polymer substrates is shown in Fig. 9. While remarkable shape changes such as flattening or pseudopod formation of

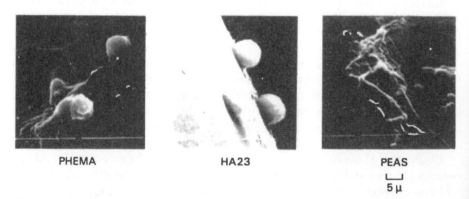

Figure 9. Morphology of attached lymphocytes.

lymphocytes were observed on the surface of poly HEMA or poly(N,N-diethyl-aminoethylstyrene) (PEAS), a model polymer of the graft chain portion in the copolymer, little change in the shape was observed on HA23, even though it retained many more lymphocytes than poly HEMA. Very little change in the shape of retained lymphocytes was also observed on the polymer surface pretreated with albumin. Further, a similar result was observed in lymphocyte adsorption on polyamine graft copolymers (SAx) with a polystyrene backbone which has a more hydrophobic nature than HAx. The results indicate that prevention of cellular shape change is one of the common characteristics of polyamine graft copolymers having microphase-separated structures, and suggest their surfaces have an ability to inhibit contact induced activation of blood cells.

The effect of cellular metabolism on cell adsorption was evaluated by use of cytochalasin B (CB), which inhibited reorganization of cytoskeletal microfila-ments, or by lowering the adsorption temperature to inhibit the metabolism [30]. Even though the extent of lymphocyte retention on homopolymer surfaces was clearly reduced by pretreating lymphocytes with CB, little change in lym-phocyte retention on the HAx copolymer was shown regardless of the pretreat-ment. Thus, reorganization of cytoskeletal microfilaments was shown to play a significant role in the cell adsorption on homopolymer surfaces but not in that on the copolymer surfaces. Further, a decrease in cell retention on the homo-polymer was observed by lowering the adsorption temperature from 37 to 4°C, whereas no change or, in some cases, an increase in cell retention was observed for the copolymer with a reduction in temperature, revealing that the polyamine graft copolymers adsorb blood cells independent on their metabolism. In other words, cell adsorption on the polyamine graft copolymers is suggested to be determined primarily by the first process in Fig. 6, where only the physico-chemical interactions between cellular surfaces and polymer surfaces are the driving force of the adsorption.

3.3. Separation of lymphocyte subpopulations by polyamine graft copolymers

From section 3.2, one can expect that polyamine graft copolymers would dif-ferentially adsorb cell populations with physicochemically different natures in their cellular membranes. As rat mesenteric lymphnode lymphocytes consist of two major subpopulations (B cells and T cells), differences in adsorption behavior between B and T cells was then examined. The capability of polymer surfaces to selectively adsorb B cells, or T cells, from a mixture was quantified by the use of the value A_B/A_T as the selectivity index, which is induced from the following relationship [31].

$$-\log([B]/[B]_0) = A_B \lambda \tag{1}$$

$$-\log([T]/[T]_0) = A_T \lambda \tag{2}$$

$$-\log([B]/[B]_0) = A_B/A_T[-\log([T]/[T]_0)] \tag{3}$$

where, λ is column length, $[B]_0$ and $[T]_0$ are the initial concentrations of B and T cells, respectively, prior to loading into the column, $[B]$ and $[T]$ are B and

Table 2.
Percentage retention of lymphocytes and A_B/A_T values obtained for polymer colums pretreated with albumin

Column	A_B/A_T[a]	Percentage retention[b]
PHEMA	1.71 (14)	36.3 ± 3.6
HA7	ND	1.3 ± 0.7
HA13	5.05 (18)	25.7 ± 3.1
HA23	2.73 (16)	57.1 ± 3.0
PEAS	1.11 (8)	94.4 ± 0.9
nylon	1.70 (6)	29.6 ± 1.7

[a] The number of data points are shown in parentheses.
[b] Mean ± SEM.

T cell concentrations, respectively, in the suspension effused from the column with a correction for column void volume. From eqns (1) and (2), eqn (3) can be obtained, and from the value of the ratio A_B/A_T we can quantitatively evaluate the selectivity of the polymer substrate to adsorb a particular subpopulation from a mixture. Namely, higher values of A_B/A_T indicate preferential adsorption of B cells over T cells. Values of A_B/A_T obtained for albumin-pretreated polymer substrates examined are summarized in Table 2 [32]. While a value of 1.7 for A_B/A_T was obtained for poly HEMA, indicating its poor selectivity, a value over 5.0 was found for HA13, indicating highly selective adsorption of B cells over T cells. Even HA 23, which shows higher value of percentage retention than PHEMA, exhibits an A_B/A_T value of 2.7. It is to be noted that polyamine graft copolymers with a polystyrene backbone (SA copolymers) also showed a higher selectivity than polystyrene in the presence of albumin pretreatment, revealing a significant role of polyamine grafts for selective adsorption of B cells over T cells. Of interest, HA copolymers behaved differently from SA copolymers in the selective adsorption of lymphocyte subpopulations in the absence of albumin-pretreatment. Although SA copolymers lost their selectivity toward B cells in the absence of albumin pretreatment, the HA copolymers retained their excellent selectivity even in the absence of albumin. SA copolymers strongly interact with lymphocytes in the absence of albumin due to the hydrophobic interaction between polystyrene domains and lymphocytes. The strong hydrophobic interaction induced nonselective adsorption of B and T cells. The hydrophobic interaction would be reduced by the adsorbed albumin layer, resulting in preferential adsorption of B cells.

In contrast to SA copolymers, HA copolymers are composed of relatively hydrophilic domains of poly HEMA and polyamine domains, so that no strong hydrophobic interaction took place between the surfaces and lymphocytes even in the absence of albumin, resulting in selective adsorption of B cells regardless of albumin pretreatment. Results of the practical separation of lymphocyte subpopulations by the use of a column packed with 1 g of HA coated glass beads with or without albumin pretreatment are summarized in Table 3. Regardless of pretreatment with albumin, a T cell suspension with 90% purity was revealed

Table 3.
Separation of B and T cells by the HA13 column with or without albumin-pretreatment[a]

Run	Percentage of B cells in initial cell population	Albumin pretreatment	Percentage of B cells in effluent cell population	Percentage recovery	
				total	T cells
1	26.3	+	10.2	56.8	69.2
		−	7.8	51.1	64.0
2	28.1	+	11.3	60.4	74.6
		−	7.6	50.7	65.1
3	26.9	+	10.7	48.5	61.4
		−	8.7	40.1	53.3
4	29.9	+	8.6	52.0	67.8
		−	6.6	40.0	53.4

[a] Separations were carried out at 23°C. 1×10^7 cells were passed through the column packed with 1 g of beads.

to be obtained as column effusion with yield over 60%. Cell sorter analyses of initial and effluent suspensions further showed the successful separation by HA copolymers as shown in Fig. 10. In addition, separation properties of HA copolymers were found to be further improved by controlling the separation conditions, such as the initial concentration of the lymphocyte suspension and loading rate as well as by regulating polymer structure, such as its molecular weight and its distribution of polyamine graft chains [33, 34].

Separation conditions of polyamine graft copolymers for lymphocyte sub-population, compared with that of the nylon fiber column (NF) method are summarized in Table 4. The NF method essentially requires incubation of

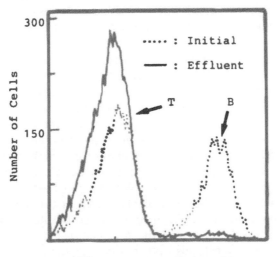

Figure 10. FACS analysis for initial and effluent cell populations. Separations were carried out at 4°C using a HA13 column.

Table 4.
Conditions for lymphocyte separation by nylon fiber, SA- and HA-columns.

Polymer	Temperature (°C)	Serum component	Time required (min)
nylon fiber	37	FCS[a]	30–60 (with incubation)
SA50	4–37	albumin	5
HA13	4–37	not required	5

[a] Fetal calf serum.

lymphocytes in the column in the presence of fetal calf serum for 30–60 min at 37°C. Actually, columns packed with glass beads coated with nylon 6 showed almost no resolution toward lymphocyte subpopulations. This indicates that the NF method separates lymphocyte subpopulations based on differences in active adhesion (process II) which was enhanced by the incubation at 37°C. In comparison to the NF method, separation by polyamine graft copolymer column does not require incubation of lymphocytes in the column, and is completed within 5 min regardless of separation temperatures ranging from 4 to 37°C. These characteristics of polyamine graft copolymers strongly indicate that these copolymers discriminate lymphocyte subpopulations through differences in their passive adsorption, recognizing difference in physicochemical properties of the plasma membrane surface of subpopulations. This consideration is further shown by the fact that cells separated by the copolymer column retained excellent viability (more than 95%).

Mechanisms involved in the selective adsorption of lymphocyte subpopulations were elucidated with particular attention to the ionic nature of polyamine graft copolymers, because ionic interactions between the negatively charged cell surface and the basic polymer surface is considered to be one of the major driving forces of lymphocyte adsorption. As ionic interaction was affected by the surrounding ionic strength and pH, the effect of the factors on lymphocyte adsorption on HA copolymers was evaluated. Lymphocyte adsorption on HA copolymers increases with decreasing the ionic strength of the cell suspension, indicating ionic interaction working as a major adsorption force. Lymphocyte adsorption on the copolymers was also increased with decreasing the pH from 8.0 to 5.5. As the zeta potential of lymphocyte membrane surface is reported to be unchanged in this pH range, the change in lymphocyte retention with varying pH is considered to be due to a change in the ionic nature of the copolymers. In fact, the degree (α) of protonation of the amino groups in the copolymer was found to be remarkably varied in this pH range. Namely, ionic interaction between acidic groups on the cellular membrane surface and protonated amino groups on the polymer surface is suggested to be a driving force for lymphocyte adsorption. Actually, the extent of lymphocyte or platelet adsorption was shown to be well correlated with the amount of protonated amino groups regardless of the difference in the amount of amino groups in the copolymers. Thus, adsorption of B and T cells on HA copolymers

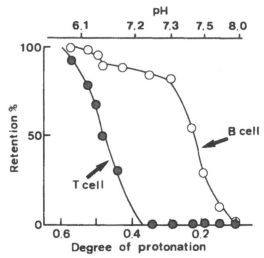

Figure 11. Changes in B and T cell retention with protonation degree (α) of the polyamine graft.

was evaluated in relation to the degree of protonation (α) of the amino groups [35].

Figure 11 shows relationships between α and the percentage retention of B and T cells. It is to be noted that the B cell retention increased steeply with increasing α compared to that of T cells, indicating B cells were more liable to adsorb on the polymer surface with protonated amino groups than T cells. This result coincided with the result of electric focusing measurements, which indicated the isoelectric points of B and T cells to be 3.8 and 4.6, respectively [36]. B cells with higher acidic groups interacted more strongly with the polymer surface than with protonated amino groups. Thus, polyamine graft copolymers were revealed to discriminate lymphocyte subpopulations based on differences in their surface ionic nature. Further, the conformation of polyamine graft chains on the polymer surface is one of the important factors which affect separation profiles of polyamine graft copolymers.

3.4. Polyamine-grafted nylon 6 as a leukocyte separator

There is another type of multiphase polymer with polyamine grafts showing high utility in the field of cell separation. That is polyamine-grafted nylon 6 utilized for granulocyte separation [37]. A major drawback of granulocyte separation from the whole blood based on the adhesion of granulocytes to a fiber- or bead-shaped adsorbent is the poor recovery of the attached granulocyte from the adsorbent surface [38]. By using our polyamine-grafted nylon 6 as an adsorbent, this drawback was remarkably improved to allow 70% recovery of adhered granulocytes from the copolymer column (Fig. 12). This high recovery of granulocytes from the polyamine-grafted nylon 6 column indicates the essentially reversible nature of the attached granulocytes due to their reduced contact-induced activation.

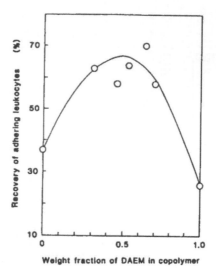

Figure 12. Recovery of leukocytes from polyamine–grafted nylon 6 columns [37].

3.5. Multiphase polymers with polypeptide microdomains

A series of microdomain-structured adsorbents containing hydrophobic synthetic polypeptide, poly (γ-benzyl L-glutamate) (PBLG), as one of the micro-domains, were prepared, and their feasibility to resolute lymphocyte sub-populations was examined [39, 40]. An important facet of these studies was to mimic biospecific interactions of natural polypeptides which participate in the molecular recognition in biological entities. Properties of the partner domains linked with PBLG were varied from hydrophobic to hydrophilic. The molecular structures of the prepared copolymers are summarized in Fig. 13. As is the case for microdomain-structured surfaces composed of neutral and cationic parts like the PHEMA–polyamine graft copolymer, these multiphase-polypeptide surfaces have the property to suppress morphological changes in adhered lym-phocytes. No such suppression was observed for lymphocytes attached on homogeneous polypeptide surfaces.

Another unique feature is that preferential adsorption of B cells over T cells was always observed for all of these multiphase-polypeptide surfaces so far examined. As no ionic groups exist on these surfaces, this resolution between B and T cells seems to be based on the differential hydrophobic interaction. Thus, in turn, surface hydrophobicity as well as the surface ionic character of B and T cells seems to differ to a considerable extent.

Acknowledgements

This research was supported by the Ministry of Education, Science and Culture, Japan (Special Project Research, Design of Multiphase Biomedical Materials), and was done in cooperation with Professors S. Inoue, University of Tokyo, N. Ogata, Sophia University and I. Shinohara, Waseda University.

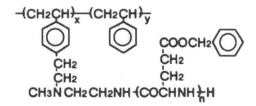

Polystyrene-PBLG graft copolymer

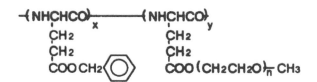

PBLG-PEG graft copolymer

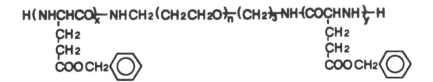

PBLG-PEG block copolymer

Figure 13. Molecular structures of block and graft polypeptides.

REFERENCES

1. Singer, S. J. and Nicolson, G. L., The fluid mosaic model of the structure of cell membrane. *Science* **175**, 720 (1980).
2. Nicolson, G. L., Transmembrane control of the receptors on normal and tumor cells. I. Cytoplasmic influence over cell surface components. *Biochim. Biophys. Acta* **457**, 57 (1976).
3. Kataoka, K., Control of cell adhesion with polymeric materials. In: *Biomaterials Science, Vol.* 1, Eds Tsuruta, T. and Sakurai, Y., Nankodo, Tokyo, 1982, p. 93.

4. Kataoka, K., Okano, T., Akaike T., Sakurai, Y., Maeda, M., Nishimura, T., Nitadori, Y., Tsuruta, T., Shimada, M. and Shinohara, I., Preparation and evaluation of new polymeric membranes for cell separation and cell culture. *Jpn. J. Artif. Organs* **8**, 804 (1979).
5. Kataoka, K., Sakurai, Y. and Tsuruta T., Microphase separated polymer surfaces for separation of B and T lymphocytes. *Makromol. Chem.* (Suppl.) **9**, 53 (1985).
6. Sakurai, Y., Akaike, T., Kataoka, K. and Okano, T., Interfacial phenomena in biomaterials chemistry. In: *Biomedical Polymers*, Eds Nakajima, A. and Goldberg, E., Academic Press, New York, 1980, pp. 335–379.
7. Okano, T., Aoyagi, T., Kataoka, K., Abe, K., Sakurai, Y., Shimada, M. and Shinohara, I., The hydrophilic–hydrophobic microdomain surface having the ability to suppress platelet adhesion and their *in vivo* antithrombogenecity. *J. Biomed. Mater. Res.* **20**, 919 (1986).
8. Okano, T., Nishiyama, S., Shinohara, I., Akaike, T., Sakurai, Y., Kataoka, K. and Tsuruta, T., Effect of hydrophilic and hydrophobic microdomains on mode of interaction between block polymer and blood platelets. *J. Biomed. Mater. Res* **15**, 393 (1981).
9. Okano, T., Kataoka, K., Sakurai, Y., Shimada, M., Akaike, T. and Shinohara, I., Molecular design of block and graft copolymers having the ability to suppress platelet adhesion. *Artificial Organs* (Suppl.) **5**, 468 (1981).
10. Shimada, M., Miyahara, M., Tahara, H., Shinohara, I., Okano, T., Kataoka, K. and Sakurai, Y., Synthesis of 2-hydroxyethyl methacrylate–dimethylsiloxane block copolymers and their ability to suppress blood platelet aggregation. *Polymer J.* **15**, 649 (1983).
11. Okano, T., Uruno, M., Sugiyama, N., Shimada, M., Shinohara, I., Kataoka, K. and Sakurai, Y., Suppression of platelet activity on microdomain surfaces of 2-hydroxyethyl methacrylate–polyether block copolymers. *J. Biomed. Mater. Res.* **20**, 1035 (1986).
12. Okano, T., Katayama, M. and Shinohara, I., The influence of hydrophilic and hydrophobic domains on the water wettability of 2-hydroxyethyl methacrylate–styrene copolymers. *J. Appl. Polymer Sci.* **22**, 369 (1978).
13. Okano, T., Nishiyama, S., Shinohara, I., Akaike, T. and Sakurai, Y., Interaction between plasma protein and microphase separated structure of copolymers. *Polymer J.* **10**, 223 (1978).
14. Okano, T., Kataoka, K., Abe, K., Sakurai, Y., Shimada, M. and Shinohara, I., *In vivo* antithrombogenecity of block copolymers having hydrophilic–hydrophobic microdomains evaluated by atriovenous shunts method. *Prog. Artif. Organs*, Vol. 2 Eds Atsumi, K., Maekawa, M. and Ota, K., ISAO Press, Cleveland, 1983, p. 863.
15. Joyce, L. D., DeVries, M. C., Hastings, W. L., Olsen, U. B., Jurvik, R. K. and Kolff, W. J., Response of the human body to the first permanent implant of the Jarvik-7 total artificial heart. *Trans. Am. Soc. Artif. Intern. Organs* **29**, 81 (1983).
16. Yui, N., Tanaka, J., Sanui, K., Ogata, N., Kataoka, K., Okano, T. and Sakurai, Y., Characterization of the microstructure of poly(propylene oxide)-segmented polyamide and its suppression of platelet adhesion. *Polymer J.* **16**, 119 (1984).
17. Yui, N., Sanui, K., Ogata, N., Kataoka, K., Okano, T. and Sakurai, Y., Effect of microstructure of poly(propylene oxide)-segmented polyamides on platelet adhesion. *J. Biomed. Mater. Res.* **20**, 929 (1986).
18. Yui, N., Kataoka, K., Sakurai, Y., Sanui, K., Ogata, N., Takahara, A. and Kajiyama, T., ESCA study of new antithrombogenic materials: surface composition of poly(propylene oxide)-segmented nylon 610 and its blood compatibility. *Makromol. Chem.* **187**, 943 (1986).
19. Yui, N., Kataoka, K. and Sakurai, Y., Microdomain-structured polymers as antithrombogenic materials. In: *Artificial Heart I*, Eds Akutsu, T., Koyanagi, H., Pennington, D. G., Poirier, V. L., Takatani, S. and Kataoka, K., Springer, Tokyo, 1986, pp. 23–30.
20. Kataoka, K., Okano, T., Sakurai, Y., Nishimura, T., Maeda, M., Inoue, S. and Tsuruta, T., Effect of microphase separated structure of polystyrene/polyamine graft copolymer on adhering rat platelets *In Vitro. Biomaterials* **3**, 237 (1982).
21. Kataoka, K., Okano, T., Sakurai, Y., Nishimura, T., Inoue, S., Watanabe, T. and Tsuruta, T., Adhesion behavior of lymphnode lymphocytes on polystyrene/polyamine graft copolymer surfaces having microphase separated structures. *Makromol. Chem., Rapid Commun.* **3**, 275 (1982).
22. Watanabe, T., Maruyama, A., Yoshino, N., Tsuruta, T., Kataoka, K., Okano, T., Sakurai, Y., Nishimura, T. and Inoue, S., Control of B and T lymphocyte adhesion by polystyrene/polyamine comb-type copolymer having microphase separated structures. *Jpn. J. Artif. Organs* **11**, 1171 (1982).
23. Folkman, J and Moscona, A., Role of cell shape in growth control. *Nature* **273**, 345–347 (1978).
24. Welpert, L., Macpherson, I. and Todd, I., Cell spreading and cell movement; an active or a passive process. *Nature* **223**, 512–513 (1969).

25. Nitadori, Y. and Tsuruta, T., Lithium amide catalyzed anionic polyaddition of 1,4-divinylbenzene with *N,N'*-diethylethylenediamine. *Makromol. Chem.* **180**, 1877 (1979).
26. Nishimura, T., Maeda, M., Nitadori, N. and Tsuruta, T., Synthesis and copolymerization of a polyamine macromer: molecular design of a new functional graft copolymer. *Makromol. Chem.* **183**, 29 (1982).
27. Maruyama, A., Senda, E., Tsuruta, T. and Kataoka, K., Synthesis and characterization of polyamine graft copolymers with a poly(2-hydroxethyl methacrylate) backbone. *Makromol. Chem.* **187**, 1895 (1986).
28. Nishimura, T., Maeda, M., Nitadori, Y. and Tsuruta, T., Polystyrene–polyamine comb-type graft copolymer: synthesis and microphase structure. *Makromol. Chem., Rapid Commun.* **1**, 573 (1980).
29. Maruyama, A., Tsuruta, T., Kataoka, K. and Sakurai, Y., Quantitative evaluation of rat lymphocyte adsorption on microdomain structured surfaces of poly(2-hydroxyethyl methacrylate)/polyamine graft copolymer by adsorption chromatography. *Biomaterials*, **9**, 471 (1988).
30. Maruyama, A., Tsuruta, T., Kataoka, K. and Sakurai, Y., Elimination of cellular active adhesion on microdomain structured surface of polyamine graft copolymers. *Biomaterials*, in press.
31. Kataoka, K., Okano, T., Sakurai, Y., Nishimura, T., Inoue, S., Watanabe, T., Maruyama, A. and Tsuruta, T., Differential retention of lymphocyte subpopulations (B and T cells) on the microphase separated surface of polystyrene/polyamine graft copolymers. *Eur. Polym. J.* **19**, 979 (1983).
32. Maruyama, A., Tsuruta, T., Kataoka, K. and Sakurai, Y., Separation of B- and T-lymphocytes by cellular adsorption chromatography using poly(2-hydroxyethyl methacrylate)/polyamine graft copolymer as column adsorbent. *J. Biomed. Mater. Res.* **22**, 555 (1988).
33. Nabeshima, Y., Maruyama, A., Tsuruta, T. and Kataoka, K., A polyamine macromonomer having controlled molecular weight—synthesis and mechanism. *Polymer J.* **19**, 593 (1987).
34. Nabeshima, Y., Tsuruta, T., Kataoka, K. and Sakurai, Y., Structural control of poly(2-hydroxyethyl methacrylate)-graft-polyamine copolymers for differential retention of rat lymphocyte subpopulations. *J. Biomater. Sci. Polym. Ed.*, in press.
35. Maruyama, A., Tsuruta, T., Kataoka, K. and Sakurai, Y., Polyamine graft copolymer for separation of rat B and T lymphocytes. Role of ionic interaction between polymer matrix and lymphocytes. *Makromol. Chem., Rapid Commun.* **8**, 27 (1987).
36. Hirsh, R. L. and Gray, I., Fractionation of rat blood lymphocytes by isoelectric focusing. *J. Immunol. Methods* **18**, 95–104 (1977).
37. Yui, N., Sanui, K., Ogata, N., Kataoka, K., Okano, T. and Sakurai, Y., Reversibility of granulocyte adhesion using polyamine-grafted nylon-6 as a new column substrate for granulocyte separation. *Biomaterials* **6**, 709 (1985).
38. Kataoka, K., Polymers for cell separation. In: *Critical Reviews in Biocompatibility*, **4**, 341 (1988).
39. Maeda, M., Kimura, M., Inoue, S., Kataoka, K., Okano, T. and Sakurai, Y., Adhesion behavior of rat lymphocyte subpopulations (B cell and T cell) on the surface of polystyrene/polypeptide graft copolymer. *J. Biomed. Mater. Res.* **20**, 25 (1986).
40. Yokoyama, M., Nakahashi, T., Nishimura, T., Maeda, M., Inoue, S., Kataoka, K. and Sakurai, Y., Adhesion behavior of rat lymphocytes to poly(ether)–poly(amino acid) block and graft copolymers. *J. Biomed. Mater. Res.* **20**, 867 (1986).

Multiphase Biomedical Materials, pp. 21–40 (1989)
T. Tsuruta and A. Nakajima (Eds)
© 1989 VSP.

Chapter 2

Microdomain structure and biomedical properties of block copolymers containing polypeptide blocks

AKIO NAKAJIMA,[1] TOSHIO HAYASHI[2] and HIROKO SATO[2]

[1]*Department of Applied Chemistry, Osaka Institute of Technology, Ohmiya, Asahi-ku, Osaka 535, Japan*
[2]*Research Center for medical Polymers and Biomaterials, Kyoto University, Sakyo-ku, Kyoto 606, Japan*

Summary—This article surveys the authors' investigations on microphase-separated membrane-forming synthetic block copolymers containing polypeptide blocks, as materials with potential towards their biomedical applications in surgical prosthetic devices. Hydrophobic and amphipathic types of *ABA* block copolymers are dealt with. Theoretical considerations are made on the formation of novel microheterophase structures as revealed by peptide–elastomer–peptide type block copolymers, the mechanical and dynamic mechanical behaviors of the membranes, and on the surface characterization of, and plasma protein adsorption on, the block copolymer membranes. In the light of information on the molecular and cellular levels, some properties required for biomedical applications, such as antithrombogenicity, biocompatibility and biodegradation, are discussed on the basis of *in vitro* and *in vivo* tests.

1. INTRODUCTION

Since Sadron's pioneering works [1, 2] on 'organized structures' produced by block copolymers, investigations on the formation and structures of micro-domains formed by solvent casting from solution of block copolymers of *AB* and *ABA* types have been extensively advanced. Hitherto, even domain structures as revealed by rather complicated block copolymers as three-component penta-block, *ABACA*, copolymers have been well clarified [3].

In recent years, synthetic block or segmental copolymer materials exhibiting microheterophase structures have been interesting in biomedical applications from the view point of biomimetic analogy; all the cells and tissues in the living body are evidently built up from microheterophase structures. The most abundant polymeric materials in the living body are obviously proteins. In this sense, the use of synthetic polymers containing protein-like blocks, such as polypeptide blocks and polyurethane blocks as biomedical materials would be expected to be useful in respect of biocompatibility. As pointed out by Anderson [4], considerations on antigenicity are also important when poly amino acids are used for injection or implantation into mammals. However, materials which may show antigenicity are not to be dealt with here.

Not only purely synthetic block copolymers, but also hybrid block copolymers composed of naturally-occurring biopolymer block and synthetic polypeptide block, are interesting. Douy and Gallot [5] have obtained a hydrid block copolymer by polymerizing an amino acid to a biopolymer Oβ extracted from egg white, and indicated the formation of a well-defined microheterophase structure [3].

In this chapter, domain structures and biomedical properties of novel synthetic block copolymers containing polypeptide blocks will be described mainly from fundamental standpoints of view, based on authors' research work.

2. MICROPHASE STRUCTURES AND MECHANICAL PROPERTIES

2.1. ABA-*type block copolymers whose middle block is an elastomer and side blocks are polypeptides*

Various kinds of *ABA* block copolymers carrying polypeptides as the *A* component and an elastomer as the *B* component were synthesized. Those are designated as GBG, MBM, EBE, LBL, MTM and ETE, in which G, M, E and L, respectively, indicate poly(γ-benzyl L-glutamate), poly(γ-methyl L-glutamate), poly(γ-ethyl L-glutamate) and poly(ε-*N*-benzyloxycarbonyl L-lysine), and B and T, respectively, indicate polybutadiene and poly(tetramethylene oxide). Homopolymers are designated as PBLG for poly(γ-benzyl L-glutamate) for an example.

The formation and structures of microdomains produced by block copolymers were quantitatively studied by Kawai [6–8], Meier [9, 10], and others for *AB*- and *ABA*-type block copolymers whose both block component chains are non-ionic and which exist in random coil conformation in solution. However, the block copolymers we are concerned with fall in to a different category from theirs, because the polypeptide chain portion of the block copolymers such as GBG, MTM, etc., exists in an α-helical conformation in helicogenic solvent as well as in the cast membrane.

From thermodynamic considerations on micelle formation, as shown in Fig. 1, for such 'rod-coil–rod' type block copolymers, we have derived expressions [11] to obtain the equilibrium micelle dimensions, $D_{s,eq}$, $D_{c,eq}$ and L_{eq}, respectively, for spherical, cylindrical and lamella-like micelles.

$$D_{s,eq} = (8\, \gamma_{AB} \langle r_{B/2}^2 \rangle / kTN)^{1/3} \tag{1}$$

$$D_{c,eq} = (16\, \gamma_{AB} \langle r_{B/2}^2 \rangle / 3\phi_B^{1/2} kTN)^{1/3} \tag{2}$$

$$L_{eq} = (8\, \gamma_{AB} \langle r_{B/2}^2 \rangle / 3\phi_B^2 kTN)^{1/3}, \tag{3}$$

where γ_{AB} is the interfacial tension between *A* and *B* components, $\langle r_{B/2}^2 \rangle$ is the root-mean-square end-to-end distance of the *B* chain having a degree of polymerization of $P_B/2$, ϕ_B is the volume fraction of the *B* block relative to the total volume occupied by the copolymer in solution, *N* is the number of junction

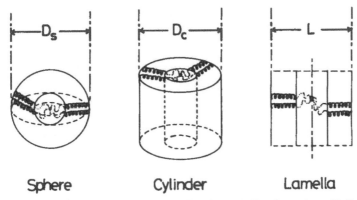

Figure 1. Model for micelle (spherical, cylindrical and lamella-like) formation of helix–coil–helix type block copolymer from solution.

points between A and B blocks per unit volume of micelle, k is the Boltzman constant and T is the temperature.

In Table 1, the equilibrium micelle dimensions calculated from eqns (1)–(3) are compared with the micelle dimensions D_{EM} estimated from electron micrographs for cast membranes [12–15]. Agreement between the observed and the calculated values seems satisfactory.

Table 1.
Micelle dimensions (Å) of ABA block copolymers in which A is polypeptide and B is polybutadiene of $P_B = 64$[a]

Code	ϕ_B	P_A	$D_{s,eq}$	$D_{c,eq}$	L_{eq}	D_{EM}	(Micelle)
GBG-12	0.603	50			309	300	(lamella)
GBG-14	0.355	78		394		400	(cylinder)
GBG-4	0.288	132		437		450	(cylinder)
GBG-6	0.216	260	447			430	(sphere)
GBG-8	0.138	580	927				(sphere)
MBM-11	0.695	40			203		(lamella)
MBM-12	0.646	65			277	300	(lamella)
MBM-13	0.489	124		322		370	(cylinder)
MBM-14	0.280	305		426		450	(cylinder)
EBE-70	0.88	14			204	260	(lamella)
EBE-40	0.68	47			264	350	(lamella)
EBE-20	0.44	127		341		380	(cylinder)
EBE-10	0.28	260		429		420	(cylinder)
EBE-05	0.17	495	470			480	(sphere)
LBL-1	0.653	38			257	350	(lamella)
LBL-2	0.557	42			301	380	(lamella)
LBL-3	0.350	98		357		400	(cylinder
LBL-4	0.250	158		422		460	(cylinder)
LBL-5	0.189	226	421			400	(cylinder)

[a] P_A and P_B denote the degrees of polymerization of A and B block chains, respectively.

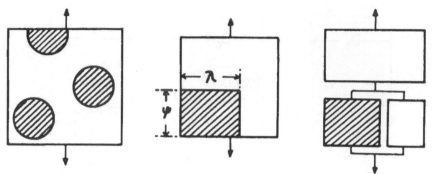

Figure 2. Equivalent mechanical model for microdomain structure of *ABA* block copolymer.

Dynamic mechanical relaxation behavior for such novel types of block copolymer membranes was examined, and the dynamic Young's modulus E^* was analyzed [16–18] by means of Takayanagi's equivalent mechanical model made up of elements connected partly in series and partly in parallel, as shown in Fig. 2, in which black portions denote the micelles of the *B* component. The relative magnitude of λ to ψ is interpreted as the ratio of the parallel to the series character, and the product $\lambda\psi$ is equal to the volume fraction of the *B* domains in the membrane. It was pointed out [14, 17, 18] that the model adopted is applicable to our block copolymers; in particular, the results obtained for block copolymer species exhibiting sphere or sphere-like microdomains quantitatively support the informations derived from eqns (1)–(3) and electron micrographs.

2.2. ABA-*type amphipathic block copolypeptides whose middle block is a hydrophobic polypeptide*

Crosslinked amphipathic block copolypeptides were prepared [19–23] by carrying out aminoalcoholysis and crosslinking reactions at the same time in the mixture of 2-amino-1-alkyl alcohol and 2.0 mol.% 1,8-octamethylene diamine (OMDA) for membranes of *ABA* block copolypeptides. The parent *ABA* block copolypeptides are GLG and GAG, where L and A denote L-leucine and L-alanine, respectively, and G indicates γ-benzyl L-glutamate as cited before. Random copolypeptides designated as GL and GA, and homopolymers designated as PBLG, PLLeu and PLAla were also synthesized as the reference materials. The parent polymers used are summarized in Table 2.

From IR spectra and wide-angle X-ray diffraction patterns (see Fig. 3 [22] as an example) for parent polymers, we pointed out [21, 22] that both G and A (or L) chains in GAG (or GLG) block copolymer exhibit microdomain structure, contrary, G and A (or L) monomeric units in GA (or GL) random copolymers cocrystallize and result in isomeric structure.

The parent polypeptides were transformed into amphipathetic polymers by partial or complete aminoalcoholysis of γ-benzyl L-glutamate (BLG) residues of the G portion into *N*-hydroxyalkyl L-glutamine (HAlkG) residues. The alkyl alcohols used for aminoalcoholysis were ethyl (E), propyl (P) and pentyl

Table 2.
Parent polypeptides for preparing amphipathic block and random copolymers

Code	BLG (mol.%)	M_w[a]
Block copolymers		
GAG-1	74.3	370,600
GAG-2	43.0	517,500
GLG-1	84.8	61,000
GLG-2	66.8	68,000
GLG-3	49.8	80,500
Random copolymers		
GA-1	75.0	348,000
GA-2	49.8	370,200
GL-1	85.0	138,000
GL-2	68.5	317,000
GL-3	48.5	312,000
Homopolymer (for reference)		
PBLG-1	100.0	371,000

[a] Weight-average molecular weight M_w determined by ultracentrifuge.

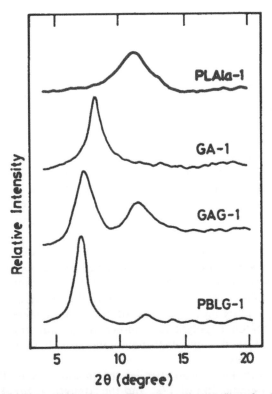

Figure 3. Wide-angle X-ray diffraction profiles of unoriented films for PBLG and PLAla homopolymers, and GAG block and GA random copolymers.

(Pe) aminoalcohols. The samples obtained are designated as GAG(E), GAG(P), GAG(Pe), GA(E), PBLG(E), etc. Formation of the crosslinked amphipathic block copolypeptides obeys the following scheme:

$$
\begin{array}{c}
\text{—(NH—CH—CO)——(NH—CH—CO)}_{\overline{m}}\text{—} \\
\quad\quad | \quad\quad\quad\quad\quad | \\
\quad (CH_2)_2 \quad\quad\quad R \\
\quad\quad | \\
\quad C{=}O \\
\quad\quad | \\
\quad OCH_2{-}\langle\bigcirc\rangle
\end{array}
\qquad
R: \begin{cases} -CH_3 \;(A) \\[2pt] -CH_2{-}CH\begin{array}{c}CH_3 \\ \\ CH_3\end{array}\;(L)\end{cases}
$$

$$\Big| \; NH_2{-}(CH_2)_x{-}OH$$

$$\text{OMDA}$$

↓

$$
\begin{array}{c}
\text{—(NH—CH—CO)———(NH—CH—CO)}_{\overline{n-1}}\text{—(NH—CH—CO)}_{\overline{m}}\text{—} \\
\quad\quad | \quad\quad\quad\quad\quad\quad | \quad\quad\quad\quad\quad\quad | \\
\quad (CH_2)_2 \quad\quad\quad (CH_2)_2 \quad\quad\quad\quad R \\
\quad\quad | \quad\quad\quad\quad\quad\quad | \\
\quad C{=}O \quad\quad\quad\quad C{=}O \\
\quad\quad | \quad\quad\quad\quad\quad\quad | \\
\quad NH \quad\quad\quad\quad NH{-}(CH_2)_x{-}OH \\
\quad\quad | \\
\quad (CH_2)_8 \\
\quad\quad | \\
\quad NH \quad\quad\quad\quad\quad NH{-}(CH_2)_x{-}OH \\
\quad\quad | \quad\quad\quad\quad\quad\quad | \\
\quad C{=}O \quad\quad\quad\quad\quad C{=}O \\
\quad\quad | \quad\quad\quad\quad\quad\quad | \\
\quad (CH_2)_2 \quad\quad\quad\quad (CH_2)_2 \quad\quad\quad R \\
\quad\quad | \quad\quad\quad\quad\quad\quad | \quad\quad\quad\quad\quad | \\
\text{—(NH—CH—CO)——(NH—CH—CO)}_{\overline{n-1}}\text{—(NH—CH—CO)}_{\overline{m}}\text{—}
\end{array}
$$

In the course of such a reaction, the alanine (or leucine) block portions are unchanged. Thus we obtain crosslinked block copolypeptides whose middle block takes an α-helical conformation, whereas side blocks take a random coil conformation, in aqueous solution.

With these samples, we confirmed that the swelling behavior in the pseudo-extracellular fluid (PECF) was well supported by Flory's swelling theory [23] based on rubber elasticity. The degrees of swelling Q_w of the membranes in PECF are plotted against the extent of partial substitution of the G-portion as represented by the OH content (mol.%) in the whole polymer, in Fig. 4(a) and (b) for L-alanine- and L-leucine-containing polymer systems, respectively. As obvious from Fig. 4(a), with increasing alkyl chain length ($x = 2$, 3 and 5), Q_w at a given OH mol.% decreases. The Q_w values of GAG(Alk) block copolymers are remarkably higher than those of corresponding PBLG(Alk) and GA(Alk) random copolymers in the low OH mol.% region. The same behavior is also found for the GLG(E) series as compared with the GL(E) series. Further, the shape of the Q_w vs. OH mol.% curves for block copolymers is quite different from those for random copolymers as shown in Fig. 4(a) and (b). Such characteristic features of block copolymer membranes are attributed to their micro-heterophase structures.

Concerning mechanical properties of the membranes in PECF, Young's modulus E at an elongation of 1%, and the strength at break point σ_B were examined as functions of Q_w. Figure 5 demonstrates a comparison of E for

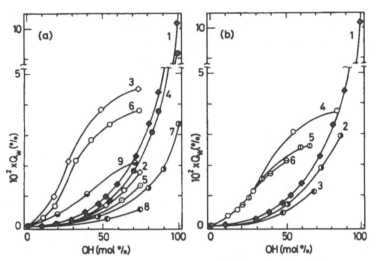

Figure 4. Degree of swelling Q_w (%) in PECF at 37°C plotted against OH content (mol.%) of polymer membrane. (a) 1: PBLG-1(E) (◆), 2: GA-1(E) (◈), 3: GAG-1(E) (◇), 4: PBLG-1(P) (●), 5: GA-1(P) (◒), 6: GAG-1(P) (○), 7: PBLG-1(Pe) (◐), 8: GA-1(Pe) (◑), 9: GAG-1(Pe) (◓). (b) 1: PBLG-1(E) (◆), 2: GL-1(E) (◑), 3: GL-2(E) (◐), 4: GLG-1(E) (○), 5: GLG-2(E) (◍), 6: GLG-3(E) (◒).

GLG(E) block copolymers with GL(E) random copolymers as functions of Q_w. Obviously, GLG(E) block copolymer membranes maintain desirable mechanical properties in PECF, for such applications as artificial skin substitutes, in a much broader region than PBLG(E) and GL(E) random copolymer membranes, i.e. GLG(E) block copolymers give softer membranes in PECF than GL(E)

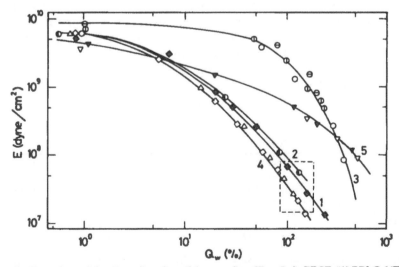

Figure 5. Young's modulus E as a function of degree of swelling Q_w in PECF. (1) PBLG-1(E) (◆), (2) GL-2(E) (◐), (3) GLG-1(E) (○), GLG-2(E) (◍) and GLG-3(E) (◒), (4) GAG-1(E) (◇) and GAG-2(E) (△), (5) GBG-1(E) (▼) and GBG-2(E) (▽).

random copolymers at the same order of Q_w. The preferable condition for artificial skin substitutes is indicated by an area surrounded by dotted lines. Further, it is pointed out that the hydrophobic nature of the glutamine side chain affects the mechanical properties of membranes, and the longer the side chain, the higher the values of E and σ_B obtained.

3. SURFACE CHARACTERIZATION

Apart from the bulk morphology, surface properties of membranes having microheterophase structures are important when discussing interactions of biological molecules and cells to the materials for biomedical uses.

Surface characterizations on our *ABA* block copolymer membranes have been carried out [24–26] by using X-ray photoelectron spectroscopy (XPS), attenuated total reflection infrared (ATR-IR) spectroscopy, replication electron micrography and contact angle measurements. The XPS method gives information on the surface layer of a few tens of angstroms in depth, so that it complements information obtained from contact angle measurements on the outermost surface and from replication electron micrography. The depth of IR beam penetration of ATR-IR spectroscopy is considerably longer and of the order of a few thousands of angstroms [25], if corresponding wave numbers are higher than 1000 cm^{-1}.

It should be noted that the surface characteristics are definitely dependent on the casting conditions of the membranes, such as the casting solvent, casting temperature and casting atmosphere. Needless to say, if a membrane is cast on a glass plate in the air, the air-facing surface of the membrane is different from the glass-facing surface, as quantitatively shown by ATR-IR [25].

When using XPS for investigating the surface composition, some devices, such as the introduction of some element of high electron density, must be taken into consideration depending on the chemical composition of the block copolymers. For example, for LBL block copolymers (see Section 2.1.), osmium tetraoxide (OsO_4) was reacted with olefinic C=C double bonds, and the osmium 4f spectra of the adduct obtained were analyzed to determine the surface content of butadiene.

Table 3.
XPS analysis for LBL-5[a]/OsO_4 adduct and its argon-etched products

Sample	Binding energy (eV)			Area ratio Os $4f_{7/2}$/C 1s	Surface content of butadiene (mol.%)
	Os(VI)	Os(IV)	Os(II + O)		
LBL-5/OsO_4	55.1	52.6	—	0.19	31
LBL-5/OsO_4 etched for 10 min	—	52.0	50.8	0.13	19
LBL-5/OsO_4 etched for 20 min	—	52.4	51.0	0.09	12
PCBL/OsO_4	—	52.4	—	0	0

[a] LBL-5: Butadiene content in the bulk = 11.9 mol.%.

(1) LBL-5   (2) PCBL

Figure 6. Replication electron micrographs of LBL-5 and PCBL surfaces.

A series of LBL block copolymers were cast from $10:1$ (v/v) mixtures of chloroform and trifluoro ethanol onto glass plates at 25°C and relative humidity less than 65%. On the air-facing LBL membrane, OsO_4 was reacted. For such an adduct and its products etched by argon, the values for the binding energy for osmium $4f_{7/2}$ peaks, peak area ratio and surface content of butadiene are shown in Table 3. The Roman numerals in the brackets indicate the valency of osmium. The carbon 1s core level was divided into three peaks, among which a peak at 287.8 eV was assigned to carbonyl carbons. It is pointed out that the outermost surface of LBL-5 contains 21 mol.% butadiene, but by argon etching of the surface, the butadiene content is reduced and finally comes to the value 11.9 mol.% for the bulk composition of LBL-5.

Figure 6 shows replication electron micrographs of membranes of LBL-5 and PCBL (homopolymer). The surface of PCBL is almost smooth, whereas that of LBL-5 contains convex circular domains of B dispersed in the planar matrix phase of L. The critical surface tensions of both the L component (41 dyn/cm) and the B component (31 dyn/cm) are larger than that of the casting solvent (25 dyn/cm), so that LBL is negatively adsorbed in the solution surface, and in the course of casting (concentration) the B component, having lower critical surface tension than the L component, becomes dominant in the outermost surface. Such an explanation is not inconsistent with the fact that the butadiene content is higher at the surface than in the bulk.

4. ADSORPTION OF PLASMA PROTEINS ONTO BLOCK COPOLYMER MEMBRANE SURFACE

When a synthetic polymer material is introduced into the cardiovascular system, rapid adsorption of plasma proteins takes place as an initial event in a complex

A. Nakajima et al.

series of reactions. The protein adsorption process leading to the equilibrium may be treated by assuming two rate-determining steps. The first step is the diffusion-controlled adsorption of protein molecules from the solution onto the material surface, for which Fick's second law is applied. As adsorption proceeds, the interfacial tension is reduced, and, accordingly, the interfacial pressure is raised. So that, in the second step, the adsorption rate is controlled by the energy barrier to be overcome. The energy barrier to be crossed over to create a space of area A in a surface film of surface pressure Π in order to adsorb a molecule would be ΠA.

Taking these two steps into consideration, we have derived the following equation for time-dependent protein adsorption from solution onto polymer membrane surface.

$$\frac{d\Pi}{dt} = k_1 \left\{ c_0 \left(\frac{d\Gamma}{d\Pi} \right)^{-1} - \frac{1}{\sqrt{\pi D}} \int_0^t \frac{\Pi'(\tau)}{\sqrt{t-\tau}} d\tau \right\} \exp\left(-\frac{\Pi A}{kT} \right) - k_2 \Pi, \tag{4}$$

where c_0 is the initial concentration of protein in the solution, D the diffusion constant of the protein, Γ the number of protein molecules adsorbed per unit area, t the time, $\pi = 3.142$ and k the Boltzman constant, k_1 and k_2 are the rate constants for adsorption and desorption, respectively, and Π' indicates $d\Pi/dt$.

In Fig. 7, $\log(d\Pi/dt)$ values obtained by numerical calculation from eqn. (4) for irreversible adsorption (i.e. $k_2 = 0$) are plotted against Π for various c_0 values. These curves distinctly indicate that the diffusion-controlled step is rapidly transferred to the energy barrier-controlled step which is shown by the

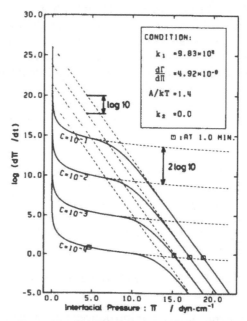

Figure 7. Dependence of calculated $\log(d\Pi/dt)$ on interfacial pressure Π for various initial concentration c_0.

broken lines with steeper slopes. From experimentally obtained $\log(d\Pi/dt)$ vs. Π curves, we can estimate the A value.

Experimentally, the adsorption of bovine serum albumin BSA and of bovine serum γ-globulin IgG at $c_0 = 0.001$ (w/w) in PECF (composed of $NaHCO_3$, K_2HPO_4, NaCl, KCl and water, $I = 0.128$ and pH 7.4) onto membrane surfaces of GLG block copolymers, GLG(E) amphipathic block copolymers and GL random copolymers has been examined by using eqn. (4). The interfacial pressure $\gamma_{sw'}(t)$ of protein solution (w') on the polymer surface (s) as a function of t was experimentally obtained from eqn. (5) by measuring the interfacial tensions γ_{Hw} and $\gamma_{Hw'}(t)$, and contact angles θ and $\theta'(t)$.

$$\Pi_{sw'}(t) = \gamma_{Hw} \cos \theta + \gamma_{Hw'}(t) \cos \theta'(t), \tag{5}$$

where γ_{Hw} and $\gamma_{Hw'}(t)$ are the interfacial tensions between n-hexane and PECF, and between n-hexane and protein solution, respectively, and θ and $\theta'(t)$ are the contact angles of PECF and of protein solution, respectively, on polymer membrane kept in n-hexane.

Characteristics of the polymer samples used are shown in Table 4, in which elemental ratios in the surface phase were obtained from carbon ls spectra in XPS analysis. Obviously, the contact angle increases with increasing leucine content, whereby the number ratio $-O-/C$ considerably decreases. If we compare the block copolymer GLG with the random copolymer GL at the same leucine content (10.7 mol.%), the contact angle of GL-34 is considerably larger, and the ratio $-O-/C$ is considerably smaller, than that of GLG-1.

The experimentally obtained $\log(d\Pi_{sw'}/dt)$ values for adsorption of BSA and of IgG onto polymer materials were plotted against $\Pi_{sw'}$. Though the curves are not shown here, the curves obtained for the adsorption of BSA onto GLG, GL and PBLG were similar to those represented in Fig. 7, i.e. both the diffusion-controlled step at the early stage, and the energy barrier-controlled step at the later stage were observed. It was also found that the slopes of all the

Table 4.
Bulk composition and surface characteristics of GLG, GLG(E), GL and PBLG

Code	L-Leu content (mol.%)	Contact angle of PECF (deg)	Surface elemental ratio(%) obtained from C ls spectra in XPS analysis		
			O/C	=O/C	$-O-/C$
PBLG	0	118.8 ± 0.53	25.0	15.6	9.4
GLG-1	10.7	115.4 ± 0.66	26.0	16.4	9.7
GLG-2	19.1	128.1 ± 0.69	24.1	16.5	7.6
GLG-3	24.9	134.5 ± 0.54	22.9	16.1	6.8
GLG-4	40.0	137.3 ± 0.77	21.6	16.1	5.5
GL-32	5.3	123.3 ± 0.86	23.5	15.2	8.3
GL-34	10.6	139.9 ± 0.89	21.5	15.8	5.7
GL-38	21.0	144.5 ± 0.96	20.9	15.4	5.5
GLG-3(2E)	24.9	120.3 ± 1.86			
GLG-3(4E)	24.9	112.2 ± 1.06			

Table 5.
Surface pressure $\Pi_{sw'}$ and effective surface areas A of BSA and IgG on PBLG, GLG and GLG(E) surfaces

Code	L-Leu (mol.%)	γ-BLG (mol.%)	$\Pi_{sw'}$ (dyne/cm)		A (Å²)	
			BSA	IgC	BSA	IgC
PBLG	0	100.0	13.1 ± 0.12	7.1 ± 0.23	600	200
GLG-1	10.7	89.3	9.5 ± 0.20	4.8 ± 0.18	560	210
GLG-2	19.1	80.9	13.3 ± 0.10	6.9 ± 0.10	550	190
GLG-3	24.9	75.1	15.7 ± 0.14	9.3 ± 0.04	620	190
GLG-4	40.0	60.0	18.1 ± 0.05	11.6 ± 0.12	580	190
GL-32	5.3	94.7	12.4 ± 0.13	5.2 ± 0.07	540	210
GL-34	10.6	89.4	19.3 ± 0.05	12.7 ± 0.08	560	190
GL-38	21.0	79.0	21.7 ± 0.06	15.2 ± 0.06	600	190
GLG-3(2E)	24.9		9.9 ± 0.63	9.7 ± 0.51	650	230
GLG-3(4E)	24.9		4.4 ± 0.42	10.9 ± 0.48	800	470

straight line portions are the same irrespective of GLG, GL and PBLG. On the other hand, for adsorption of IgG, the diffusion-controlled step was not distinguished in the experimental time-scale. The slopes of the straight lines for the GLG, GL and PBLG series are almost the same, but those for the GLG(E) series increase with increasing hydrophilicity. Such a difference in adsorption behavior at the early stage may be attributed to the fact that IgG is a gluco-protein whereas BSA is a simple protein. By means of the slopes of straight lines, the A values were estimated from eqn. (4). The interfacial pressure $\Pi_{sw'}$ at after 3 h from the beginning of the experiments, and the effective cross-sectional areas A of BSA and IgG, for all the polymer materials tested are summarized in Table 5.

As obvious from Table 5, the $\Pi_{sw'}$ values of BSA and IgG for GL random copolymers are higher than the respective values for GLG block copolymers of the same leucin content. This means that the GL surface is more hydrophobic than the GLG surface of the same bulk composition. The A values of BSA (~ 600 Å²) for both GLG and GL series are about three times those of IgG (~ 200 Å²) and $\Pi_{sw'} A$ for BSA is always higher than that for IgG. This means that IgG is more easily adsorbed in the energy barrier-controlled process. For the GLG-3(E) series, derived from GLG-3 by aminoalcoholysis for 2 h [designated as (2E)] and for 4 h (4E) with 2-amino-1-ethanol, the adsorbed amount of IgG in the early stage is larger than that of BSA, and the A value of IgG (470 Å²) was larger than that of the parent GLG-3 (190 Å²). From these findings, we suggest that the F_{ab} portion of γ-globulin orients toward the hydrophilic surface by an end-on fashion, in contrast to the orientation of F_c portion by end-on fashion toward the hydrophobic surface.

Another sophisticated method for directly observing the adsorption of plasma proteins on polymer surface is the use of immunoelectron microscopy [27, 28]. After immersing the polymer material in human blood for 2 h, the protein species adsorbed were identified by reacting respective anti-human

Table 6.
Adsorption of plasma proteins in human blood as estimated by immunoelectron microscopy

Polymer	Albumin	Fibrinogen	γ-Globulin	β-Lipoprotein
EBE-10	+	−	+	+
EBE-20	+	−	−	−
EBE-40	−	−	−	−

−: negative, +: slightly positive, ++: positive, +++: strongly positive.

serum, i.e. anti-albumin serum, anti-fibrinogen serum, anti-γ-globulin serum, and anti-β-lipoprotein serum, and the reaction products were further reacted with peroxidase-labeled anti-rabbit IgG and oxidized by 3,3'-diaminobenzidine tetrahydrochloride. Finally, the oxidized specimens were stained with OsO_4 and observed with an electron microscope [29]. Some experimental results [29, 30] on the EBE block copolymer series are shown in Table 6.

5. ANTITHROMBOGENICITY

Since Imai [31] pointed out in 1972 that polymer materials exhibiting micro-heterophase structures of some specified dimension are prominent in anti-thrombogenicity, interactions of block and segmented copolymer materials with plasma proteins and blood cells such as platelet have been extensively investigated to elucidate the modes of antithrombogenicity in connection with the microheterophase structures revealed on the material surface. Meanwhile, for example, Okano *et al.* have extended a number of works, as introduced in their review article [32], on *ABA* block copolymers composed of hydroxyethyl methacrylate as the *A* component, and dimethyl siloxane or styrene as the *B* component. They arrived [33] at an important conclusion that the inhibition effect toward adhesion and aggregation of platelets closely relates to the size of the microdomain structures, and the microdomain structures of order 10^2 Å are critically important for antithrombogenicity.

Concerning our *ABA* block copolymers, *in vivo* antithrombogenicity tests were carried out by Noishiki [28, 29] for the GBG, MBM, EBE, CBC and GTG series, whereby surgical silk sutures coated with test polymers were placed in the vein of a mongrel dog for 2 weeks, and the test specimens taken out were observed. Among block copolymers tested, antithrombogenicity of the EBE series was predominant as shown in Table 7. The domain sizes of EBE series are in the range 350–420 Å and their shapes are lamellar and cylindrical.

In view of the above-mentioned results on antithrombogenicity, Nakajima [34] proposed the following hypothesis about the interaction of platelets with block copolymer materials having microheterophase structures. When such a material is placed into blood, adsorption of plasma proteins takes place in accordance with the microdomain structures of the material, i.e. the *A* and *B* domains are predominantly covered with different kinds of plasma proteins.

Now we assume that the radius of a platelet cell is 10000 Å, and the number of glycoproteins distributed in the cell membrane is 10000, then the surface area

Table 7.
In vivo antithrombogenicity tests for EBE block copolymers

Polymer	Molecular fraction of B component	Domain size (Å) and shape	Antithrombogencity
EBE-10	0.105	420 (cylinder)	+
EBE-20	0.194	380 (cylinder)	−
EBE-40	0.395	350 (lamella)	+

−: no thrombus adheres; thrombus adhered— +: 25% ++:~50%, +++: ~100%; ++++: blood vessel blocked.

per glycoprotein is estimated as 13×10^4 Å. In other words, the average distance between adjacent two glycoproteins is estimated as about 350 Å. If we presume that platelet activation, such as release reactions and deformation of platelets, is caused by the action of lining proteins located on the inner side of the cell membrane only when the sugar chains of the adjacent glycoproteins interact simultaneously with those of the adsorbed plasma proteins such as γ-globulin and fibrinogen, then the foreign-body recognition of the platelet is not realized for domain dimension of about 350 Å (see Fig. 8). This figure of 350 Å is in good agreement with the domain size for preferable antithrombogenicity as pointed out from Table 7.

Together with such dimensional considerations, considerations on hydrophilicity and hydrophobicity of the *A* and *B* domains are also important. As indicated in Table 5, the surface pressure for BSA is higher for hydrophobic surfaces than for hydrophilic surfaces. In such sense, amphipathic block copolymers, such as GLG(E), GAG(E), etc., of some specified hydrophilicity would also be useful candidates as antithrombogenic biomaterials.

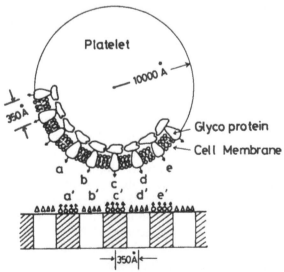

Figure 8. Interaction of platelet cell with protein-adsorbed surface of microdomain structure.

6. BIOCOMPATIBILITY

In vivo biocompatibility tests for block copolymers were carried out by Noishiki, whereby block copolymer samples coated on polyester fiber mesh cloth were implanted into the back muscle of rabbit for 4 weeks. The samples taken out together with the tissue surrounding the polymer sample were fixed in 10% formalin and incubated in paraffin. The sliced specimens were stained with hematoxylin and eosin, and then examined microscopically. The results obtained [15, 30, 35, 36] are summarized in Table 8.

Although some remarkable foreign-body reactions were observed with GBG-11, MBM-14 and LBL-15, the foreign-body reaction and adsorbance by the living body for other samples are considerably low. In particular, the EBE series seems to be promising for biomedical applications in view of its antithrombogenicity and biocompatibility. A photomicrograph showing the situation of EBE-20 in surrounding tissue is given in Fig. 9 as an example.

In contrast to these rather rigid copolymer materials (Table 8), the amphipathic block copolypeptides as indicated in Section 2.2 are rather elastomeric, and were confirmed by *in vivo* tests to display excellent tissue compatibility. *In vivo* tests were carried out [37–39] by implanting the specimens into the back muscle of rabbit, and evaluations of these materials were made for such applications as temporary barriers to prevent adhesion between natural tissue planes [37, 38] and temporary skin substitutes [39]. The block copolymers of the

Table 8.
In vivo biocompatibility test for various block copolymers

Polymer	Molecular fraction of *B* component	Foreign-body reaction	Absorbance by living body
GBG-11	0.508	+ +	+ +
MBM-12	0.433	±	−
MBM-14	0.091	+ + +	±
EBE-05	0.055	±	−
EBE-10	0.105	+	− or ±
EBE-20	0.194	±	− or ±
EBE-40	0.395	+	−
LBL-12	0.423	+	−
LBL-15	0.185	+ + +	+
GTG-3	0.200	+	
MTM-3	0.132	±	
ETE-3	0.132	±	

Foreign-body reaction:
   −: no reaction, +: inbetween − and +,
   +: light reaction, + +: inbetween + and + +,
   + + +: remarkable reaction, + + + +: necrosis of cells.
Degree of adsorbance:
   −: no adsorbance, +: inbetween − and +,
   +: ~25% test mater. is adsorbed, + +: ~50%,
   + + +: ~75%, + + + +: >95%.

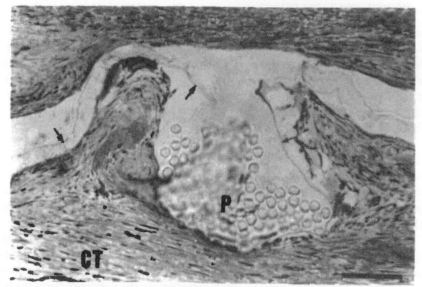

Figure 9. Microphotograph of EBE-20-coated fibers (P) and surrounding tissue. CT and arrow indicate connective tissue and EBE-20, respectively. Bar: 100 μm.

GAG(E), GAG(P), GLG(E) and GBG(P) series were found to have predominant biocompatibility and to maintain mechanical strength as a temporary barrier material to prevent adhesion.

For temporary skin substitute application, permeation of water vapor through the membrane is also important as well as tissue compatibility. Rate

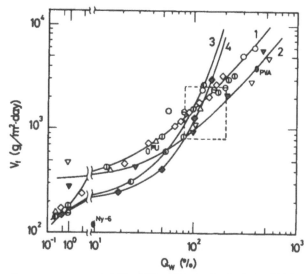

Figure 10. Rate of water vapor permeability V_f through membrane from PECF, plotted against degree of swelling Q_w. (1) GLG-1(E) (O), GLG-2(E) (Φ), GLG-3(E) (⊖), GAG-1(E) (◇) and GAG-2(E) (△), (2) GBG-1(E) (▼) and GBG-2(E) (▽), (3) GL-2(E) (◉) and (4) PBLG-1(E) (◆).

of water vapor permeability V_f from PECF through a polymer membrane is shown in Fig. 10, in which desirable region is indicated by an area surrounded by dotted lines (cf. V_f of normal human skin is 300–500 g/m^2 day).

7. BIODEGRADATION

Poly(α-amino acid)s and their copolymers are typical biodegradable polymers. For this reason, biomedical applications as mentioned in the latter half of Section 6 are most adequate. For simulative purpose, *in vitro* biodegradation of amphipathic copolypeptides, as indicated in Section 2.2, was studied [21, 22] by using papain, an endopeptidase, and pronase E, an exopeptidase.

Figures 11 and 12 illustrate the weight change of GL(E), GLG(E) and

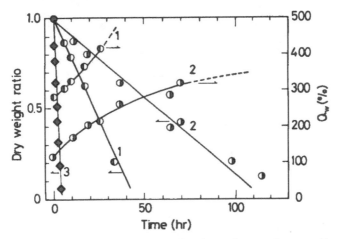

Figure 11. Dry weight ratio and Q_w plotted against time for random copolymers and homopolymer. (1) GL-1(E) (◑), (2) GL-2(E) (◐) and (3) PBLG-1(E) (◆).

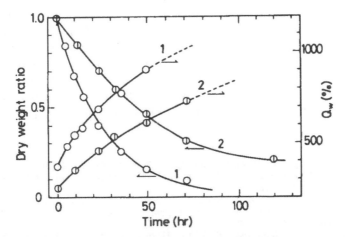

Figure 12. Dry weight ratio and Q_w plotted against time for block copolymers. (1) GLG-1(E) (○) and (2) GLG-2(E) (◑).

PBLG(E) membranes (crosslinked with 2.0 mol.% OMDA) as functions of digestion time in 0.1 mg/ml pronase solution in PECF (pH 7.4) at 37°C. All the samples were obtained from parent polymers (see Table 2) by converting all the γ-benzyl L-glutamate residues into *N*-hydroxyethyl L-glutamine residues. As obvious from Figs 11 and 12, dry weight ratio vs. digestion time curves are almost linear for random copolymers and homopolymers, but concave above for block copolymers reflecting the domain structure.

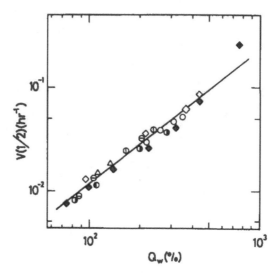

Figure 13. Rate of pronase E digestion $V(1/2)$ of membranes plotted against degree of swelling Q_w in PECF at 37°C and pH = 7.4 for PBLG-1(E) (◆), GL-2(E) (◑), GLG-1(E) (○), GLG-2(E) (◐), GLG-3(E) (⊖), GAG-1(E) (◇) and GAG-2(E) (△).

In Fig. 13, the rates of pronase E digestion $V(1/2)$, defined as the reciprocal of the time required for the sample weight to be reduced to one-half its initial value, are plotted against the degree of swelling. A similar curve was also obtained for papain digestion. These results indicate that the degree of swelling of the sample in water is a unique measure for the rate of digestion, irrespective of homopolymer, random copolymer, and block copolymer. Results of these model experiments *in vitro* have to be correlated to *in vivo* biodegradation and adsorption in living body. Some *in vivo* tests are under way [37–39].

REFERENCES

1. Sadron, C., Organized polymers. *Pure Appl. Chem.* **4**, 347–352 (1962).
2. Sadron, C., Ungleichmäßig lösliche makromoleküle—heterogele und heteropolymere. *Angw. Chem.* **75**, 472–475 (1963).
3. Hunabashi, H., Miyamoto, T., Isono, T., Fujimoto, T., Matsushita, Y. and Nagasawa, M., Preparation and characterization of a pentablock copolymer of the *ABACA* type. *Macromolecules* **16**, 1–4 (1983).
4. Anderson, J. M., Spilizewski, K. I. and Hiltner, A., Poly-α-amino acids as biomedical polymers. In: *Biocompatibility of Tissue Analogs*, Ed. Williams, D. F., CRC Press, Boca Rabon, 1985, pp. 67–88.

5. Douy, A. and Gallet, B., Synthesis and ordered structure of amphipathic block copolymers with a saccharide and a peptide. *Biopolymers* 19, 493–507 (1980).
6. Inoue, T., Soen, T., Kawai, H., Fukatsu, M. and Kurata, M., Electron microscopic texture of *AB* type block copolymers of isoprene with styrene. *J. Polymer Sci. B* 6, 75–81 (1968).
7. Inoue, T., Soen, T., Hashimoto, T. and Kawai, H., Thermodynamic interpretation of domain structure in solvent-cast films of *AB* type block copolymers of styrene and isoprene. *J. Polymer Sci. A-2* 7, 1283–1302 (1969).
8. Uchida, T., Soen, T., Hashimoto, T. and Kawai, H., Domain structure and bulk properties of solvent-cast films of *ABA* type block copolymers of styrene–isoprene–styrene. *J. Polymer Sci. A-2* 10, 101–121 (1972).
9. Meier, D. J., Theory of block copolymers I. Domain formation in *AB* block copolymers. *J. Polymer Sci. C* 26, 81–98 (1969).
10. Meier, D. J., Theory of block copolymers. *Polymer Prepr. ACS, Div. Polym. Chem.* 11, 400–401 (1970).
11. Nakajima, A., Kugo, K. and Hayashi, T., Microheterphase structure of the *ABA* type block copolymer consisting of the α-helical poly(γ-benzyl L-glutamate) as the *A* component and polybutadiene as the *B* component. *Macromolecules* 12, 844–848 (1979).
12. Nakajima, A., Hayashi, T., Kugo, K. and Shinoda, K., Synthesis and structural study of the *ABA* type block copolymer consisting of poly(γ-benzyl L-glutamate) as the *A* component and polybutadiene as the *B* component. *Macromolecules* 12, 840–843 (1979).
13. Hayashi, T., Chen, G. W., Kugo, K. and Nakajima, A., Organized structure and properties of polyamino acid block copolymers. *Kasen Koen-Shu* (in Japanese) 38, 37–48 (1981).
14. Kugo, K., Hayashi, T. and Nakajima, A., Synthesis, structure, and mechanical properties of *ABA* tri-block copolymers consisting of poly(ε-N-benzyloxycarbonyl L-lysine) as the *A* component and polybutadiene as the *B* component. *Polymer J.* 14, 391–399 (1982).
15. Chen, G. W., Sato, H., Hayashi, T., Kugo, K., Noishiki, Y. and Nakajima, A., Microheterphase structure, permeability, and biocompatibility of *ABA* tri-block copolymer membranes composed of poly(γ-ethyl L-glutamate) as the *A* component. *Bull. Institute of Chem. Res., Kyoto University* 59, 269–283 (1981).
16. Hayashi, T., Anderson, J. M. and Hiltner, P. A., Block copolypeptides. 2. Viscoelastic properties. *Macromolecules* 10, 352–356 (1977).
17. Nakajima, A., Kugo, K. and Hayashi, T., Mechanical properties and water permeability of *ABA* tri-block copolymer membranes consisting of poly(γ-benzyl L-glutamate) as the *A* component and polybutadiene as the *B* component. *Polymer J.* 11, 995–1001 (1979).
18. Chen, G. W., Hayashi, T. and Nakajima, A., Synthesis and molecular characterization of *ABA* tri-block copolymers composed of poly(γ-ethyl L-glutamate) as the *A* component and polybutadiene as the *B* component. *Polymer J.* 13, 433–442 (1981).
19. Tabata, Y., Mochizuki, M., Minowa, K., Hayashi, T. and Nakajima, A., Synthesis and properties of copolypeptide membranes composed of *N*-hydroxyethyl glutamine as one component. *Rep. Progr. Polym. Phys., Japan* 25, 683–686 (1982).
20. Hayashi, T., Minowa, K., Takeshima, K., Tabata, Y. and Nakajima, A., Evaluation of hydrophilic block copolypeptide membranes as an artificial skin component. II. PBLG–PLL–PBLG block copolypeptide. *Polymer Preprints, Japan* 32, 2079–2082 (1983).
21. Hayashi, T., Tabata, Y., Takeshima, K. and Nakajima, A., Preparation and properties of *ABA* tri block copolymer membranes consisting of *N*-hydroxyethyl L-glutamine as the *A* component and L-leucine as the *B* component. *Polymer J.* 17, 1149–1157 (1985).
22. Hayashi, T., Takeshima, K. and Nakajima, A., Preparation and properties of *ABA* tri-block copolymer membranes consisting of *N*-hydroxyalkyl L-glutamine as the *A* component and L-alanine as the *B* component. *Polymer J.* 17, 1273–1280 (1985).
23. Flory, P. J., *Principles of Polymer Chemistry*, Cornell University Press, Ithaca, 1953.
24. Kugo, K., Hata, Y., Hayashi, T. and Nakajima, A., Studies on membrane surfaces of *ABA* tri-block copolymers consisting of poly(ε-N-benzyloxycarbonyl L-lysine) as the *A* component and polybutadiene as the *B* component. *Polymer J.* 14, 401–410 (1982).
25. Kugo, K., Murashima, M., Hayashi, T. and Nakajima, A., Structure and properties of membrane surfaces of *ABA* tri-block copolymers consisting of poly(γ-methyl D,L-glutamate) as the *A* component and polybutadiene as the *B* component. *Polymer J.* 15, 267–277 (1983).
26. Nakajima, A. and Hata, Y., Adsorption behavior of plasma proteins on polyamino acid membrane surface. *Polymer J.* 19, 493–500 (1987).
27. Nakane, P. K. and Pierce Jr, G. B., Enzyme-labeled antibodies of the light and electron microscopic localization of tissue antigens. *J. Cell. Biol.* 33, 309–318 (1967).

28. Noishiki, Y., Application of immunoperoxidase method to electron microscopic observation of plasma protein on polymer surface. *J. Biomed. Mater. Res.* **16**, 359–368 (1982).

29. Sato, H., Morimoto, H., Nakajima, A. and Noishiki, Y., Study on interactions between plasma proteins and polymer surface. *Polymer J.* **16**, 1–8 (1984).

30. Nakajima, A., Expanding scope of biomaterials in connection with antithrombogenicity. *Tutorial Lecture at the 4th Congress of International Society for Artificial Organs,* Kyoto, November 15, 1983.

31. Imai, Y., Materials carrying antithrombogenicity. *Kobunshi* (in Japanese), **21**, 569–573 (1972).

32. Okano, T., Shinoda, M., Shinohara, I., Kataoka, K., Akaike, T. and Sakurai, Y., Role of microphase separated structure in interaction between polymer and platelet. In: *Advances in Biomaterials, Vol. 3*, Eds Winter, G. D., Gibbons, D. F. and Plenk, H., John Wiley, 1982, p. 445.

33. Okano, T., Kataoka, K., Sakurai, Y., Shimada, M., Akaike, T. and Shinohara, I., Molecular design of block and graft copolymers having the stability to suppress platelet adhesion. *Artif. Organs* (Suppl.) **5**, 468 (1981).

34. Nakajima, A., Interactions in biopolymer and synthetic polymer systems. *J. Jpn. Soc. Biomaterials* (in Japanese) **2**, 67–76 (1984).

35. Noishiki, Y., Nakahara, Y., Sato, H. and Nakajima, A., Study on the relationship between biocompatibility and chemical structure of amino acid homopolymers and copolymers. *Artif. Organs* (in Japanese) **9**, 678–682 (1980).

36. Sato, H., Nakajima, A., Hayashi, T., Chen, G. W. and Noishiki, Y., Microheterophase structure, permeability, and biocompatibility of *ABA* tri-block copolymer membranes composed of poly(γ-ethyl L-glutamate) as the *A* component and polybutadiene as the *B* component. *J. Biomed. Mater. Res.* **191**, 135–1155 (1985).

37. Nakamura, T., Hitomi, S., Hayashi, T., Watanabe, S., Shimizu, Y. and Nakajima, A., Evaluation of hydrophilic polyamino acid membranes as wound-covering. *In vivo* tests for rabbit implanting. *Jpn. Soc. Biomaterials Preprints* (in Japanese) **6**, 13–14 (1984).

38. Hayashi, T., Ikada, Y., Watanabe, S., Shimizu, Y., Nakamura, T. and Hitomi, S., Evaluation of synthetic polyamino acid membranes as biodegradable absorbable materials. *J. Jpn. Soc. Biomaterials Preprints* (in Japanese) **7**, 173–175 (1985).

39. Hayashi, T., Ikada, Y., Watanabe, S., Shimizu, Y., Nakamura, T., Hitomi, S. and Nakajima, A., Preparation and membrane properties of hydrophilic poly(α-amino acid)s as biodegradable materials (3). *Polymer Preprints, Japan* (in Japanese) **34**, 1745–1748 (1985).

Multiphase Biomedical Materials, pp. 41–58 (1989)
T. Tsuruta and A. Nakajima (Eds)
© 1989 VSP.

Chapter 3

Interaction of cells on the surface of multiphase polymer assemblies

YUKIO IMANISHI

Department of Polymer Chemistry, Faculty of Engineering, Kyoto University, Yoshida Honmachi, Sakyo-ku, Kyoto 606, Japan

Summary—The long-term, *in vivo* antithrombogenicity of heparinized polyetherurethaneureas suggested that the promotion of endothelialization of synthetic materials by controlling surface properties is promising for the development of truly biocompatible materials.

To estimate properties of synthetic materials for *in vivo* pseudo-endothelialization, interactions of polyetherurethaneurea derivatives with fibroblast cells as well as with plasma proteins were investigated. Fibronectin, which is a cell adhesion protein, was easily adsorbed by the heparinized polyetherurethaneurea, but the degree of adsorption to the polymer in competition with other proteins was so low that the cell attachment to polyetherurethaneurea was decreased by heparinization. Different degrees of cell attachment onto different materials were considered due to different extents of adsorption of plasma proteins. Proliferation of fibroblast cells was suppressed on cationic polyetherurethaneureas but unaffected on other derivatives of polyetherurethaneureas. The absence of specific suppression of cell growth indicates the usefulness of heparinized polyetherurethaneureas for truly biocompatible materials by pseudo-endothelialization.

Attachment and growth of fibroblast cells on polypeptide derivatives with different wettabilities were also studied in the presence or absence of serum proteins. In the presence of serum, a peak level of cell attachment was observed for substrates with a contact angle of around 70°. However, no relationship was found between cell attachment and water contact angles of substrate polymers in the absence of serum. Ca^{2+}-dependent cell attachment was observed on hydrophobic surfaces in the absence of serum proteins, suggesting that Ca^{2+}-dependent membrane proteins function as mediator during attachment to hydrophobic surfaces. In the presence of serum proteins, it was found that cell attachment is affected by metabolism, Ca^{2+} and the cytoskeleton of the cell. Cell growth rate on hydrophilic substrates was higher than on intermediate or hydrophobic substrates, demonstrating that strong interactions between cells and substrates are unfavorable for a dewebbing process during mitosis.

1. INTRODUCTION

Recently, it has been considered most promising for the development of truly biocompatible materials that materials, which specifically activate the mechanism of living body to assimilate foreign materials and to recover from injury, are designed and synthesized. The first experimental observation that led to the above idea was made by Van Kampen *et al.* [1]. They examined the *in vivo* patency of a series of vascular prostheses made of polypeptides and

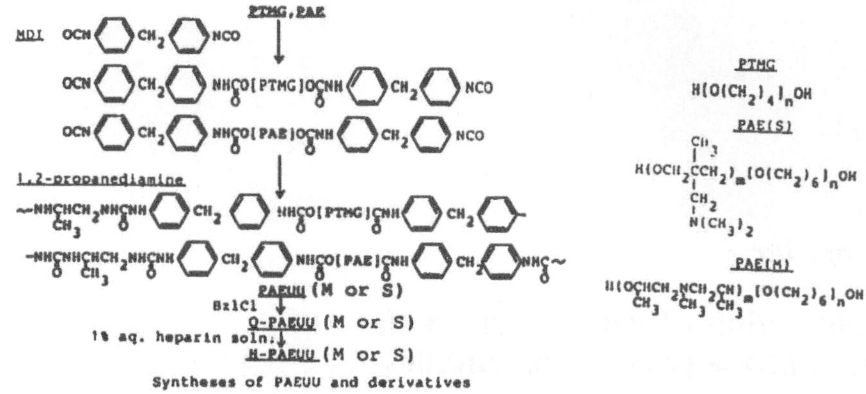

Y. Imanishi

Figure 1. Synthesis of polyetherurethaneurea to which heparin is ionically bound.

polyetherurethanes or injured blood tube, and reported that a good patency was obtained when adhesion and spreading of leukocytes were observed. Thereafter, Burkel *et al.* [2] succeeded in attaining an excellent antithrombogenicity of vascular grafts which were made of knitted Dacron velour by mixing fibrin and seeding endothelial cells.

We have succeeded in a further improvement of antithrombogenicity of polyetherurethane materials, which are relatively antithrombogenic and mechanically stable, by heparinization [3–5]. We have synthesized two kinds of heparinized polyetherurethaneureas. In one of them, heparin was ionically bound to quaternary ammonium groups in the main chain or in the substituents of polyetherurethaneureas as illustrated in Fig. 1. In the other, heparin was covalently linked to carboxyl groups in the substituents of a polyether-urethaneurea, which were produced by saponification and treatment with citric acid of ester groups in the polyetherurethaneurea synthesized by the scheme shown in Fig. 2.

Both kinds of heparinized polymers were highly antithrombogenic, and their interactions with plasma proteins and platelets were investigated [5–9]. It was found that H-PAEUU selectively adsorbs albumin without denaturation and, therefore, is not stimulating to platelets, leading to excellent antithrombogenicity. On the other hand, PEUUL-H was not favorable for the selective adsorption of albumin, caused the denaturation of plasma proteins adsorbed, and induced the adhesion and deformation of platelets, but deactivated the blood clotting system, leading to excellent antithrombogenicity. These observations indicate the importance of a multiparameter evaluation of the activation of platelets and the blood clotting system for the investigation of blood–material interactions.

Platelets were treated with cytoskeleton breakers, and the adhesion and activation of platelets on H-PAEUU were investigated in terms of the participation of cytoskeleton proteins in platelet adhesion [10]. When the surface of H-PAEUU is covered with plasma proteins, cytoskeleton proteins do not

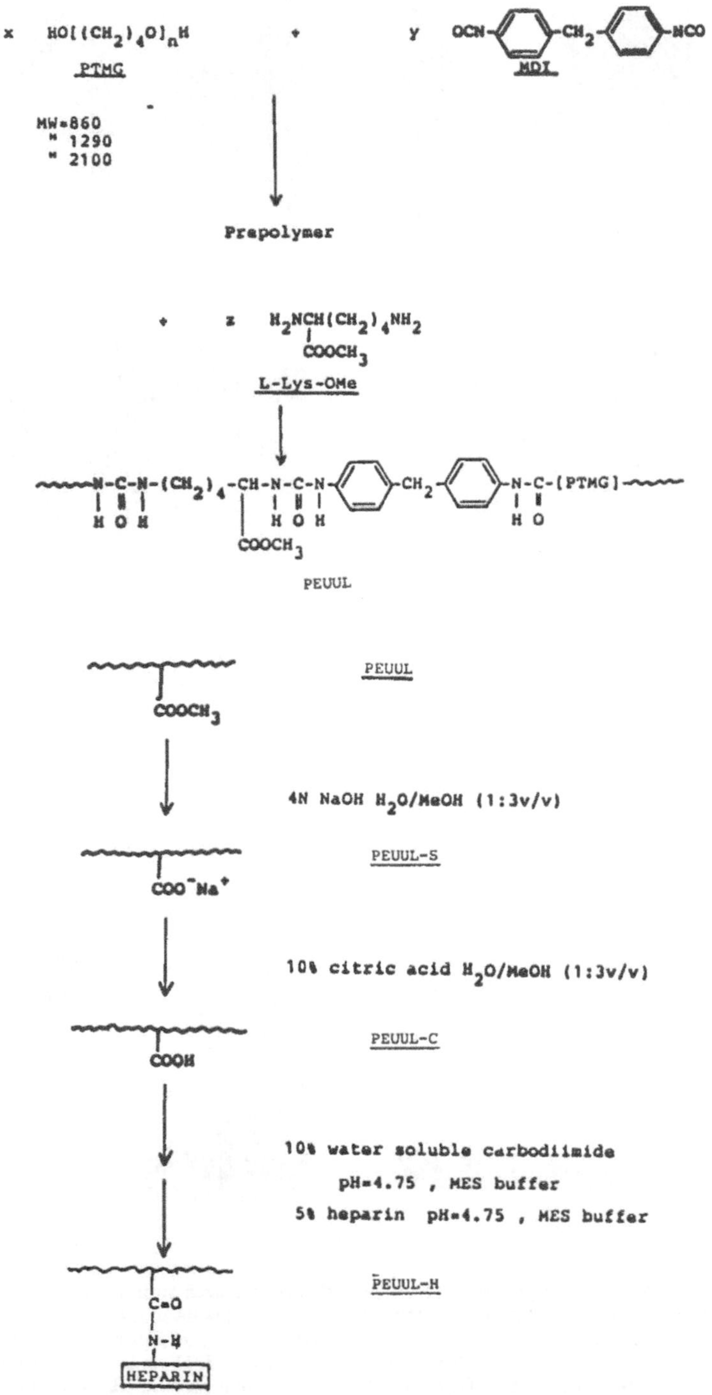

Figure 2. Synthesis of polyetherurethaneurea to which heparin is covalently bound.

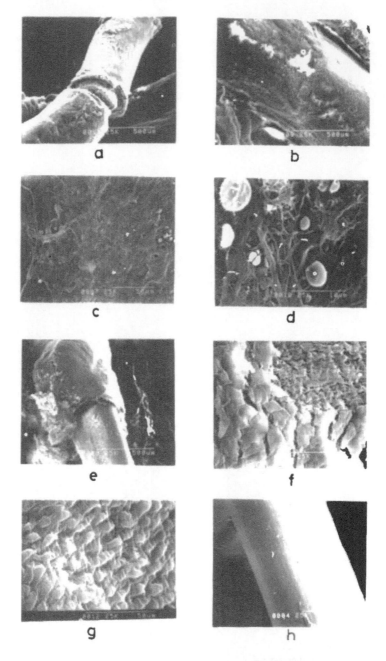

Figure 3. Scanning electron micrograph of polyetherurethanurea sutures after implantation in dogs: (a) PEUU suture covered with thrombus after 90 day implantation; (b) thrombus formed on PEUU suture near the puncture site after 90 day implantation; (c) magnification of thrombus part of (b); (d) magnification of (c); (e) thrombus formed on PEUUL-H suture near the puncture site of vein after 90 day implantation; (f) growth of endothelial cells on PEUUL-H suture after 90 day implantation; (g) magnified view of endothelial cells grown on PEUUL-H after 90 day implantation; (h) clean and smooth surface of H-PAEUU suture after 90 day implantation.

participate in the adhesion process. In other words, a passive adhesion of platelets took place on H-PAEUU. Under these circumstances, platelets were not strongly stimulated upon adhering onto H-PAEUU.

Sutures of PEUUL-H and H-PAEUU were implanted in the vein of mongrel dogs for evaluation of *in vivo* antithrombogenicity of the materials [11]. After 3 months of implantation, the H-PAEUU sutures did not produce thrombus at all on the surface and maintained a clean and smooth surface (Fig. 3h). On the other hand, PEUUL-H sutures formed thrombus a little (Fig. 3e) but endothelial cells were found to grow on the surface of the suture from the puncture point (Fig. 3f and g). Therefore, H-PAEUU is useful as an *in vivo* long-term antithrombogenic material and PEUUL-H is promising as a biocompatible implant material by pseudo-endothelialization.

With the experimental results and literature reports mentioned above in mind, adhesion and growth of mouse fibroblast cells in the presence or absence of serum proteins on polyetherurethaneureas and polypeptides with different properties were investigated to estimate properties for *in vivo* pseudo-endothelialization of the synthetic materials.

2. ATTACHMENT AND PROLIFERATION OF FIBROBLAST CELLS ON POLYETHERURETHANEUREA DERIVATIVES

2.1. Materials

Polymers employed in the present investigation are shown in Fig. 4. Ordinary polyetherurethaneurea (PEUU) was synthesized by the step-growth polymerization of poly(tetramethylene glycol) (PTMG, MW 1250), 4,4'-diphenylmethane diisocyanate (MDI) and 1,2-diaminopropane in a feeding molar ratio of $1:2:1$. Polyetherurethaneurea containing tertiary amino groups in the backbone chain (PAEUU) was synthesized by the step-growth polymerization of polyaminoether

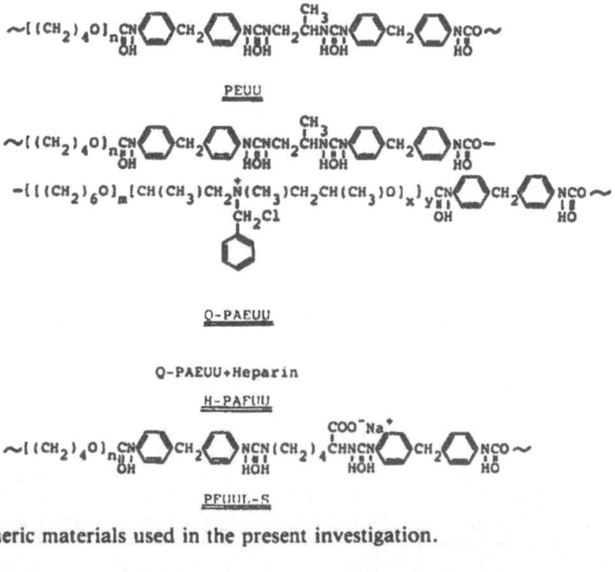

Figure 4. Polymeric materials used in the present investigation.

(PAE, MW 2180), PTMG (MW 1250), MDI and 1,2-diaminopropane in a feeding molar ratio of $1:1:4:2$. The amine content of the PAEUU was 3.45%. This polymer was completely quaternized with benzyl chloride to obtain Q-PAEUU, and heparin was bound electrostatically by Q-PAEUU to give H-PAEUU. H-PAEUU containing 21.72 wt.% heparin was used.

Polyetherurethaneurea carrying carboxylate groups as substituents (PEUUL-S) was synthesized by step-growth polymerization of PTMG, MDI and L-lysine methyl ester in a feeding molar ratio of $1:2:1$ and subsequent saponification of the ester groups.

All polymers were cast on a glass plate from 10 wt.% dimethylformamide solution. Q-PAEUU and H-PAEUU were blended with PEUU in a $1:1$ weight ratio to increase mechanical stability in wet conditions.

2.2. Cell attachment

Cultured cell STO was treated with 0.2 wt.% aqueous trypsin solution and washed with Dulbecco's modified Eagle medium (MEM). After washing, the concentrated cell suspension was treated with $Na_2{}^{51}CrO_4$ for labeling and finally suspended in fetal calf serum (FCS)-free or FCS-containing MEM, cell population being adjusted to 10^6/ml. To each of 24 well dishes in which a disinfected polymer film was placed, the cell suspension (300 ml) was added and incubation was continued under the atmosphere containing 5% CO_2 at 37°C for the requisite time. Thereafter, polymer films were washed several times with phosphate-buffered saline (PBS). ^{51}Cr was counted with an Alloka JDC-751 automatic scintillation counter.

Cells were found to attach rapidly within 1 h to all materials. Therefore, in the present investigation, cell attachment behaviors were compared at 1 h from the beginning of cell–material contact. Attachment of fibroblast cells to different materials with or without precoating by protein is compared in Table 1. In the absence of protein coating, the amount of cells attached to glass or Q-PAEUU was very high and decreased in the order of H-PAEUU > PEUU > PEUUL-S. When FCS was present, the nature of substrate polymers affected the cell attachment in nearly the same way as that in the absence of the serum. When the materials were coated with specific proteins, cell attachment was accelerated according to the nature of coating protein in the order of BSA < BγG < BPF. The effect of the substrate polymer on cell attachment was less marked than that of the coating proteins.

Table 1.
Cell adhesion on several materials

Materials	Non-coated		Protein-coated		
	without FCS	with FCS	BSA	BγG	BPF
PEUU	4.9 ± 1.0	7.7 ± 0.5	1.1 ± 0.1	1.5 ± 0.2	10.6 ± 0.8
Q-PAEUU	49.8 ± 5.6	53.9 ± 3.9	1.7 ± 0.2	2.5 ± 0.5	12.6 ± 1.4
H-PAEUU	26.2 ± 2.5	20.5 ± 2.3	1.8 ± 0.1	1.9 ± 0.3	17.3 ± 3.5
PEUUL-S	3.6 ± 0.5	1.8 ± 0.3	2.0 ± 0.3	2.1 ± 0.3	5.4 ± 0.9
Glass	50.7 ± 0.1	51.7 ± 7.5	2.3 ± 0.7	2.9 ± 0.4	41.3 ± 9.4

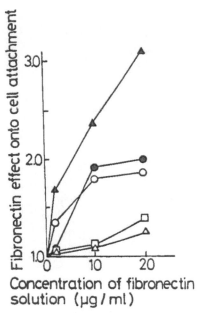

Figure 5. Effect of fibronectin on cell attachment. Cell attachment without coating by HFN was taken as unity. ●, Q-PAEUU; ○, glass; ▲, H-PAEUU; △, PEUU; □, PEUUL-S.

A change of cell attachment induced by precoating of polymer materials with fibronectin (HFN) is shown in Fig. 5. The effect of the nature of substrate polymers on cell attachment appeared more markedly than that of polymers coated with BSA, BγG or BPF. In particular, the cell attachment to H-PAEUU and Q-PAEUU was very much enhanced by coating with HFN.

Cell attachment onto H-PAEUU in the presence of serum proteins (FCS) decreased with increasing content of heparin, as shown in Fig. 6. In a

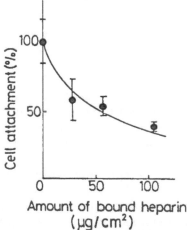

Figure 6. Effect of bound heparin on cell attachment. Cell attachment in the absence of heparin was taken as 100%.

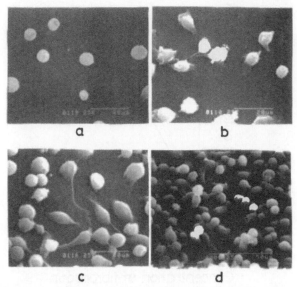

Figure 7. Scanning electron micrograph of fibroblast cells attached to different polymers: (a) PEUUL-S, (b) PEUU, (c) H-PAEUU, and (d) Q-PAEUU.

single-protein adorption experiment, HFN was shown to be adsorbed in high amounts to H-PAEUU and to promote attachment of fibroblast cells (Fig. 5). Therefore, the trend shown in Fig. 6 may have been caused by a low adsorption of HFN under competitive adsorption conditions.

Scanning electron micrographs which observe the states of cell attachment onto different materials are shown in Fig. 7. To a qualitative level it is feasible to say that the spreading of attached cells decreases according to the nature of substrate polymers and glass in the order of Q-PAEUU > glass > H-PAEUU > PEUU > PEUUL-S, which is the decreasing order of cell attachment.

Different degrees of cell attachment onto synthetic polymers have been discussed in terms of surface hydrophilicity, water content, surface charge and surface free energy. The present investigation revealed a particularly important influence of positive charge. It was also concluded that hydrophilic surfaces with anionic charges possess lower ability of cell attachment than hydrophobic surfaces. This trend is also observed in the experiments using a series of polypeptide derivatives, which are described in Section 3. These observations together indicate an important role of electrostatic interactions in cell–polymer interactions.

In the present investigation, a suppressive effect of H-PAEUU on cell attachment as compared with Q-PAEUU was clearly shown. In a different investigation with PEUUL-H, the suppression of cell attachment was also observed. On the other hand, when HFN alone is adsorbed to H-PAEUU, it undergoes structural change to promote cell attachment. These apparent inconsistencies may be explained by the consideration that in a competitive adsorption from a

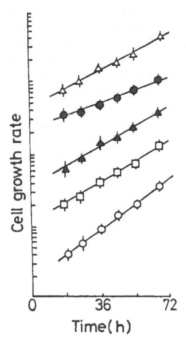

Figure 8. Cell growth rate on several materials. Rates are plotted in logarithmic scale with an arbitrary unit. △, PEUU; ●, Q-PAEUU; ▲, H-PAEUU; □, PEUUL-S; ○, glass.

protein mixture to H-PAEUU, proteins other than HFN are adsorbed preferentially. In particular, it should be noted that BSA shows a preferential adsorption to H-PAEUU.

2.3. Cell proliferation

Procedures for cell culture were nearly the same as those for cell attachment experiments, except that unlabeled cells were used. Cell culture was conducted in MEM containing 10 wt.% FCS, streptomycin and penicillin under an atmosphere containing 5% CO_2 at 37°C for the requisite time, and the number of cells was determined by the measurement of lactate dehydrogenase (LDH) activity.

As is seen in Fig. 8, cells proliferate in a logarithmic manner on all kinds of materials investigated. The rates of cell proliferation on various materials are shown in Table 2. It was found that cell proliferation was particularly suppressed on Q-PAEUU. As Imai *et al.* [12] pointed out, in the process of cell

Table 2.
Cell growth rate on several materials in arbitrary unit

Materials	PEUU	Q-PAEUU	H-PAEUU	PEUUL-S	Glass
Growth rate	0.79	0.51	0.87	0.89	1.02

proliferation, cells become initially spheric by dewebbing from substrate, then spindly, and again flat after segmentation. Therefore it may be true that dewebbing of cells from substrate is difficult, if cell–polymer interactions are strong. This idea explains the suppressive effect of Q-PAEUU on cell proliferation and receives support from scanning electron microscopic observations (Fig. 7). It also explains the mode of cell proliferations on different materials investigated, except for glass.

Heparin was found to have no particular influence on cell proliferation. It was expected that ionically bound heparin, unless released, cannot be internalized into the cytoplasm to exert biological activities. It has been recently found that free heparin accelerates adhesion of platelets but immobilized heparin suppresses it [11]. Lack of suppressive effect on cell proliferation of heparin by immobilization may affect the cell proliferation and the platelet adhesion. However, H-PAEUU did not completely inhibit cell proliferation.

3. ATTACHMENT AND GROWTH OF FIBROBLAST CELLS ON POLYPEPTIDE DERIVATIVES

A number of attempts have been made to estimate attachment and growth of cells in relation to surface properties of polymeric materials. Interactions of polypeptide with cells have been investigated in search for materials for cell culture. In this study by using synthetic polypeptides having different wettabilities, attachment and growth of fibroblast cells on these materials were investigated in relation to surface properties of the materials.

3.1. Materials
Polypeptide derivatives tested in the present investigation are listed in Table 3. They were prepared by the usual methods.

Table 3.
Materials used as substrate and their abbreviations

Sample No.	Abbreviations	Materials
1	PBLG	Poly(γ-benzyl L-glutamate)
2	PBCL	Poly(ε-N-benzyloxycarbonyl-L-lysine)
3	M-PBLG	PBLG film treated with 4N NaOH/MeOH
4	M-PBCL	PBCL film treated with HBr/ether
5	PBLG-OH	1.5 h aminolysis of PBLG film with ethanolamine
6	PBLG-90	1.5 h aminolysis of PBLG film with ethanolamine and diaminododecane
7	PBLG-20	0.3 h aminolysis of PBLG film with ethanolamine and diaminododecane
8	ABA-1	$(BLG)_m$–$(BCL)_n$–$(BLG)_m$ block copolymer
9	ABA-2	$(BCL)_m$–$(BLG)_n$–$(BCL)_m$ block copolymer
10	CO-1	PBLG containing 27% trimethylsilyl side groups
11	CO-2	PBLG containing 89% trimethylsilyl side groups
12	PS-1	PBLG containing 30% PDMS side groups
13	R-10	PBLG containing 10% fluoroalkyl side groups
14	R-30	PBLG containing 30% fluoroalkyl side groups
15	GLASS	CORNING, borosilicate glass

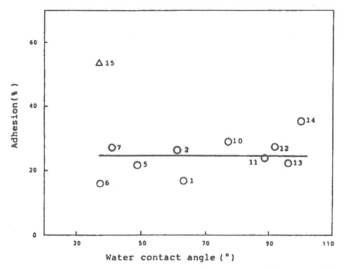

Figure 9. Mouse STO cell adhesion to polypeptide derivatives in Eagle's MEM without serum at 37°C as a function of water contact angles in air; △, glass.

3.2. Attachment of fibroblast cells (without serum)

Cell attachment experiments were carried out by the same method as described in Section 2.2. Attachment of fibroblast cells, which were suspended in MEM without serum, to polypeptide derivatives was investigated and the results are shown in Fig. 9. The ordinate represents the number of cells adhered relative to the cell number in an aliquot of the cell suspension. Cell attachment to polypeptide derivatives was not influenced by the wettability of polymer surfaces.

The results of fibroblast cell attachment to the polypeptide substrates under different conditions are shown in Fig. 10(a)–(c). When the adhesion experiment was carried out at 4°C, the amount of cell attachment was very low as compared with that at 37°C and therefore did not seem to be influenced by the wettability of substrates (Fig. 10(a)). Participation of cell metabolism in the cell attachment was suggested. When EDTA was added to the culture medium to trap Ca^{2+}, attachment of fibroblast cells was more extentive on wettable substrates than on hydrophobic substrates (Fig. 10(b)). The participation of Ca^{2+}-dependent membrane proteins in cell attachment to hydrophobic substrates was suggested. The effect of cytochalasin B addition on the cell attachment is shown in Fig. 10(c). Addition of cytochalasin B suppressed the cell attachment to the polypeptide derivatives and made the cell attachment apparently independent of the nature of the substrates. The participation of cytoskeleton proteins in the cell attachment was suggested.

3.3. Attachment of fibroblast cells (with serum)

Figure 11 summarizes the results of cell attachment to the polypeptide materials in Eagle's MEM containing 10 wt.% FCS. Cell population was high on the

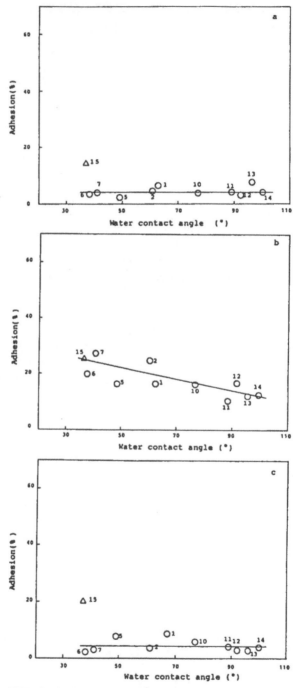

Figure 10. Mouse STO cell adhesion to polypeptide derivatives as a function of water contact angles in air; (a) in Eagle's MEM at 4°C, (b) in Eagle's MEM with EDTA at 37°C, (c) in Eagle's MEM at 37°C with the addition of cytochalasin B.

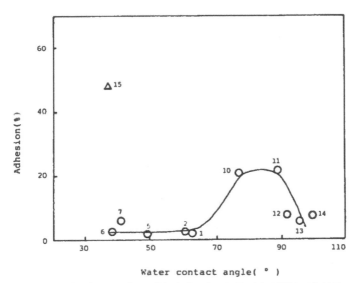

Figure 11. Mouse cell adhesion to polypeptide derivatives in Eagle's MEM with 10% serum at 37°C as a function of water contact angles in air.

substrates having water contact angles of 75–90° (Nos 10 and 11), but was very low on other substrates. This pattern of cell attachment–polymer wettability relationship is quite different from that in MEM without FCS. Since water contact angles of the polypeptide substrates treated with serum proteins are almost the same as before the treatment, the presence of the peak level of cell attachment to the polypeptide materials having water contact angles of 75–90° may be due to specific interactions between the serum proteins and the polypeptide materials.

Attachment of fibroblast cells in MEM containing FCS was investigated under different conditions and the experimental results are summarized in Fig. 12(a)–(c). Attachment of fibroblast cells was equally suppressed to a very low level by lowering temperature (Fig. 12(a)), the addition of EDTA (Fig. 12(b)) and the addition of cytochalasin B (Fig. 12(c)). Under these conditions, the peak level of cell attachment was not observed. Therefore, it is clear that specific interaction of fibroblast cells with serum proteins adsorbed on the polypeptide substrates exists and it is influenced by the cell metabolism, Ca^{2+} and cytoskeleton proteins of the cell.

3.4. Mechanism of cell attachment

The experimental results on attachment of fibroblast cells to polypeptide substrates in the presence of serum (Fig. 11) were consistent with those reported by Ikada *et al*. [13] and Tamada [14]. They reported a peak level of cell attachment on substrates having water contact angles of ~70°. However, cell attachment in the absence of serum was independent of the surface wettability of polypeptide materials (Fig. 9).

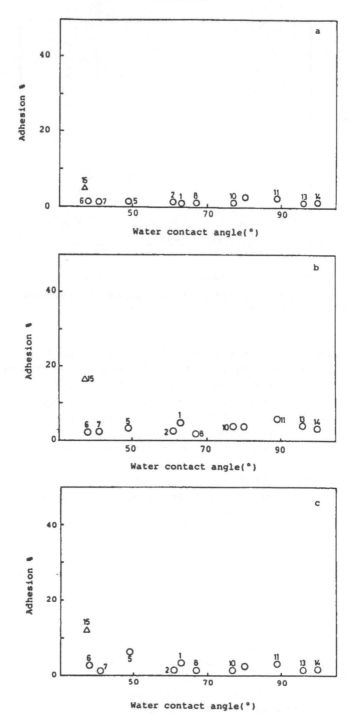

Figure 12. Mouse STO cell adhesion to polypeptide derivatives in Eagle's MEM with 10% serum as a function of water contact angles in air; (a) at 4°C; (b) with EDTA at 37°C; (c) with cytochalasin B at 37°C.

When materials were treated with serum proteins in advance, the wettability of the materials increased and became independent of the nature of the poly-peptide materials. Therefore, cell attachment on the polypeptide materials should not be affected directly by the wettability of the material surface in the presence of serum proteins. Therefore, different degrees of cell attachment according to the nature of the materials may have originated from different degrees of denaturation and different orientations of adsorbed proteins, and selective adsorption of cell-binding proteins. It could be concluded that adhesion of fibroblast cells to polypeptide derivatives is regulated by biological factors rather than by non-biological, physico-chemical factors.

In the absence of serum protein, Ca^{2+}-dependent cell attachment was observed on the hydrophobic surfaces (Fig. 10(b)). This result implies that Ca^{2+}-dependent membrane proteins function as a mediator for cells to attach to hydrophobic surfaces.

In the presence of serum proteins, it was found that cell attachment is related with metabolism in the cell (Fig. 11 and 12(a)), Ca^{2+} (Fig. 12(b)) and cyto-skeleton proteins of the cell (Fig. 12(c)). It might be argued that cells recognize proteins adsorbed and specifically oriented on the material surface having contact angles of ~70°, and undergo morphological changes of the membrane, causing patching and capping.

3.5. Growth of fibroblast cells

Procedures for cell culture were nearly the same as those for cell attachment experiments, except that unlabeled cells were used. Details are described in Section 2.3.

The number of fibroblast cells on the polypeptide substrates in the presence of serum was determined and is shown in Fig. 13 as a semi-logarithmic plot. From the slope of the straight line, the rate of cell growth was determined and is plotted against the water contact angle of the polypeptide substrates in Fig. 14. The results show that fibroblast cells grow more rapidly on more wettable substrates.

The growth rate of fibroblast cells was higher on hydrophilic substrates than on other substrates. However, cell growth, taking place via attachment and spreading, is a very complicated process. For example, the rate of cell growth is versatile according to the nature of the cell. Furthermore, it has been reported that cell growth rate decreases with increasing cell attachment [12]. Other papers report that it increases with increasing cell attachment [15, 16].

Though the patterns of cell growth were very different according to the nature of substrate, the present study showed that cells grow more rapidly under condi-tions where cell attachment is of a low level (Fig. 15).

The whole process of cell growth consists of cell adhesion, filopodial growth, cytoplasmic webbing, flattening of the cell mass, cytoplasmic dewebbing during mitosis and flattening of daughter cell. A strong interaction between cells and substrates will be unfavorable for cytoplasmic dewebbing during mitosis, thus reducing the growth rate in accordance with a previous report [12].

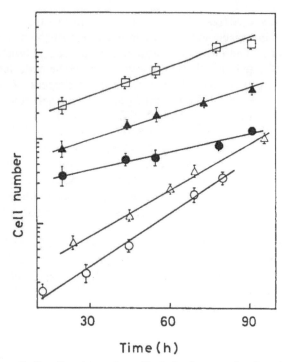

Figure 13. Number of cells adhered to polypeptide derivatives as a function of incubation time, plotted in logarithmic scale with an arbitrary unit: O, Falcon tissue culture dish; △, PBLG-90; ●, PBLG; ▲, CO-1; □, R-40.

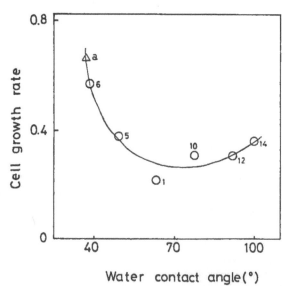

Figure 14. Cell growth rate in the presence of serum as a function of water contact angle in air, the rate on glass being taken as unity. (a) Falcon tissue culture dish.

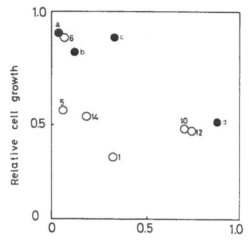

Relative cell attachment

Figure 15. Relative cell growth as a function of relative cell attachment in Eagle's MEM with 10% serum; (a) polyetherurethaneurea bearing carboxylate groups in the substituents; (b) polyetherurethaneurea; (c) quaternized and heparinized polyaminoetherurethaneurea bearing tertiary amino groups in the main chain; (d) quaternized polyaminoetherurethaneurea bearing tertiary amino groups in the main chain.

REFERENCES

1. Van Kampen, C. L., Gibbons, D. F. and Jones, R. D., Effect of implant surface chemistry upon arterial thrombosis. *J. Biomed. Mater. Res.* **13**, 517–541 (1979).
2. Burkel, W. E., Ford, J. W., Vinter, D. W., Kahn, R. H., Graham, L. M. and Stanley, J. C., Fate of knitted dacron velour grafts seeding with enzymatically derived autogolous canine endothelium. *Trans. Am. Soc. Artif. Intern. Organs* **28**, 178–184 (1983).
3. Shibuta, R., Tanaka, M., Sisido, M. and Imanishi, Y., Synthesis of novel polyaminoetherurethaneureas and development of antithrombogenic material by their chemical modifications. *J. Biomed. Mater. Res.* **20**, 971–987 (1986).
4. Ito, Y., Sisido, M. and Imanishi, Y., Synthesis and antithrombogenicity of polyetherurethaneureas containing quaternary ammonium groups in the side chains and of the polymer/heparin complex. *J. Biomed. Mater. Res.* **20**, 1017–1033 (1986).
5. Ito, Y., Sisido, M. and Imanishi, Y., Synthesis and antithrombogenicity of anionic polyurethanes and heparin-bound polyurethanes. *J. Biomed. Mater. Res.* **20**, 1157–1177 (1986).
6. Ito, Y., Sisido, M. and Imanishi, Y., Adsorption of plasma proteins to the derivatives of polyetherurethaneurea carrying tertiary amino groups in the side chains. *J. Biomed. Mater. Res.* **20**, 1139–1155 (1986).
7. Sanada, T., Ito, Y., Sisido, M. and Imanishi, Y., Adsorption of plasma proteins to the derivatives of polyaminoetherurethaneurea: The effect of hydrogen-bonding property of the material surface. *J. Biomed. Mater. Res.* **20**, 1179–1195 (1986).
8. Ito, Y., Sisido, M. and Imanishi, Y., Platelet adhesion onto protein-coated and uncoated polyetherurethaneurea having tertiary amino groups in the substituents and its derivatives. *J. Biomed. Mater. Res.*, **23**, 191–206 (1989).
9. Ito, Y., Sisido, M. and Imanishi, Y., Adsorption of plasma proteins and adherence of platelets onto novel polyetherurethaneureas—relationship between denaturation of adsorbed proteins and platelets adherence. *J. Biomed. Mater. Res*, submitted (1986).
10. Ito, Y., Imanishi, Y. and Sisido, M., Platelet adhesion onto polyetherurethaneurea derivatives: effect of cytoskeleton proteins of the platelet. *Biomaterials*, **8**, 458–463 (1987).
11. Ito, Y., Imanishi, Y. and Sisido, M., *In vitro* platelet adhesion and *in vivo* antithrombogenicity of heparinized polyetherurethaneureas. *Biomaterials*, **9**, 235–240 (1988).

12. Imai, Y., Watanabe, A., Masuhara, E. and Imai, Y., Structure–biocompatibility relationship of condensation polymers. *J. Biomed. Mater. Res.* **17**, 905–912 (1983).
13. Ikada, Y., Suzuki, M. and Tamada, Y., Polymer surfaces possessing minimal interaction with blood components. In: *Polymers as Biomaterials*, Eds Shalaby, S. W., Hoffman, A. S., Ratner, B. D. and Harbett, T. A., Plenum, New York, 1984, pp. 135–147.
14. Tamada, Y., Biomedical material surface. In: *Foundation and Application of Polymer Surface*, Ed. Ikada, Y., Kagaku Dojin, Kyoto, 1986, pp. 181–200.
15. Van Wachem, P. B., Beugeling, T., Feijen, J., Bantges, A., Detmers, J. P. and van Aken, W. G., Interaction of cultured human endothelial cells with polymeric surface of different wettabilities. *Biomaterials* **6**, 403–407 (1985).
16. Schakenraad, J. M., Busscher, H. J., Wildevuur, C. R. H. and Arends, J., The influence of substratum surface free energy on growth and spreading of human fibroblasts in the presence and absence of serum proteins. *J. Biomed. Mater. Res.* **20**, 773–784 (1986).

Multiphase Biomedical Materials, pp. 59-71 (1989)
T. Tsuruta and A. Nakajima (Eds)
1989 VSP.

Chapter 4

A new evaluation method of antithrombogenicity of biomedical polymers

YASUHARU NOISHIKI

Division of Surgery, Department of Rehabilitation Medicine, Medical School, Okayama University, Misasa, Tottori 682-02, Japan

Summary—A new *in vivo* test for the estimation of blood compatibility is demonstrated. The peripheral vein of a dog was punctured with an 18 gauge needle under general anaesthesia, and a pliable suture coated with the test polymer was inserted through the needle into the lumen of the vessel. After a given period of time, the vein in which the suture had been inserted was opened and observations were carried out macroscopically and by means of light and electron microscopy. By this method, long term interactions between the host and the test polymer in the physiological condition are easily observed.

1. INTRODUCTION

Many polymer materials with blood compatibility have been developed in expectation. They have been evaluated for their antithrombogenicity by various methods *in vivo*, *in vitro* and *ex vivo*. As screening tests, *in vitro* and *ex vivo* methods are useful to evaluate them quantitatively. The test conditions of *in vitro* methods are simpler, but less physiological. With *ex vivo* methods, the test materials can be evaluated under more physiological condition. Both methods can give exact values in each material; however, the phenomenon of the blood–material interaction is limited to observation within few minutes (*in vitro*) or few hours (*ex vivo*) after their contact. The important biological reactions such as immunological reactions, foreign body reactions and healing processes cannot be observed by such methods. These reactions require longer periods of time at physiological condition. They can be observed by *in vivo* evaluation methods.

Numerous *in vivo* methods have been developed; however, they are not very widely used for the following reasons: (i) The procedures are not easy for general investigators, requiring surgeons or experienced research workers in surgical techniques. (ii) Difficulties in the procedures may introduce errors in the evaluation. (iii) In some methods, plural materials cannot be tested in one animal, so that the control cannot be performed in the same animal. (iv) Some methods require dangerous traumatic surgical interventions to the test animals.

To overcome these difficulties, a new *in vivo* test was developed [1, 2]. The method is simple and easy to perform and reliable to estimate anti-thrombogenicity.

2. MATERIALS AND METHODS

2.1. Outline of the method

A pliable suture coated with the test polymer is inserted into the peripheral vein of an experimental animal through a syringe needle puncturing the vein. After the removal of the needle, the suture is left inside the vessel lumen. By this procedure, the test polymer can float in the peripheral venous blood at physiological conditions. The long-term relationship between the test polymer and the blood can be observed by this method. The vessel is opened after a given period of time. Observations are carried out macroscopically and by means of light and electron microscopy.

2.2. Sample preparation

A soft, pliant polyester multifilament suture, 10 cm in length and No. 2-0 (USP) in thickness, was coated with a test polymer using a dipping and solvent casting method. The coated sutures were preserved in a 70% solution of ethanol to avoid crack formation from drying and to keep them sterile. They were soaked in a sterile saline solution to remove the ethanol before implantation. In the case of soft polymer with no risk for crack formation from drying, it is possible to keep it dry and sterilize it by ethyleneoxide gas or gamma irradiation. The sample sutures had to be pliant after the coating procedure. If the suture became rigid by the coating, there was a possibility of injuring the vessel wall in which it had been inserted. In the case of such rigid sutures, it was recommended that they were precoated with a soft polymer which had an affinity for the test polymer, then the test polymer was thinly coated onto the precoated suture. Instead of a polyester multifilament suture used for the supporting structure, the test polymer itself or any other polymer having an affinity to the test polymer and having no foreign body reaction can be used as the suture material.

2.3. Animal experimentation

Adult mongrel dogs weighing 7–10 kg were anesthetized with 25 mg/kg sodium pentobarbital by intravenous injection. Endotracheal intubation was not required. Both jugular veins and both femoral veins were punctured by 18 gauge needles, percutaneously, or after open exposure of the vessels. The suture coated with the test polymer was inserted into the vessel lumen through the needle. After the needle was withdrawn, the suture was fixed to the connective tissue or skin in the vicinity of the vessel puncture to prevent it from being carried away by the blood stream (Fig. 1). The site of the vessel puncture was pressed lightly for 1–3 min to stop bleeding.

After a given period of time, while under general anesthesia, the animal was sacrificed by acute exsanguination from the aorta, with the administration of heparin (2 mg/kg). The vessel in which the suture was inserted was opened in

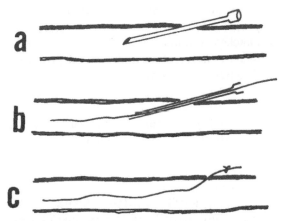

Figure 1. Procedure of the suture insertion into the peripheral vein. (a) Puncture the vein with a syringe needle; (b) insertion of the polymer coated suture through the needle; (c) fixation of the suture after the needle is withdrawn.

a longitudinal direction. During the opening, attention should be given to the tip of the scissors so as not to remove the thrombus on the suture. Observations were carried out macroscopically after washing off the blood around the suture by gently pouring on saline. A color picture was taken on the site with a name card and a scale. The suture was then removed, and fixed with 2.5% glutaraldehyde in phosphate-buffered solution for light microscopic, and scanning and transmission electron microscopic examinations. Implant time periods of 1 h, and 1, 3, 5, 7 and 14 days were used. All the test animals were autopsied to study the infarct and embolus in the lungs, the kidneys and the spleens, both macroscopically and microscopically.

2.4. Macroscopic and microscopic evaluation
The results were evaluated using the following criteria.

(1) Macroscopic observation
  Macro (−): without thrombus
     (+): partial thrombus formation
    (++): thrombus along the entire length of the suture
   (+++): obstruction of the vessel cavity by thrombus
(2) Light microscopic observation
  LMS (−): no adhesion of blood components
     (+): platelets adhesion
    (++): platelets and fibrin deposition
   (+++): thrombus deposition
(3) Scanning electron microscopic observation
  SEM (−): proteinous layer adhesion
     (+): platelets adhesion
    (++): platelets aggregation and fibrin deposition
   (+++): thrombus deposition

(4) Transmission electron microscopic observation
    TEM (−): proteineous layer adhesion
        (+): platelets adhesion
      (+ +): platelets aggregation and fibrin deposition
    (+ + +): thrombus deposition

Although each material was examined using four criteria, the results of each observation were not always so clearly defined.

2.5. Polymer materials used

Five polymers (Avcothane, H-PSD, PS-100G, SAEKA-UV and SAEKA-Hep-GA) were used in this study. Avcothane 51 elastomer [3] (registered trademark of Avco Corp.) is composed of poly(ethyl urethane) (90%) and poly(dimethylesiloxane) (10%), and was obtained from the Avco-Everett Corporation through Dr Akaike, Tokyo University of Agriculture and Technology. This polymer displays considerable hemocompatibility without any incorporated anticoagulants.

The H-PSD [4] was supplied by the Basic Research Laboratories of Toray Industries Inc., 1111 Tebiro, Kamakura, Japan. It is a recently developed heparinized hydrophilic polymer possessing heparin ionically bound within the polymer matrix. It continuously releases a certain amount of heparin from its surface when placed in the blood stream, and the rate of heparin released from the polymer surface can be controlled as desired by changing the chemical composition and water content of the hydrophilic polymer matrix [5]. The polymer that had about 30% water of adsorption by weight and approximately 15% heparin by weight showed good antithrombogenicity.

The PS-100G polymer [6], also supplied by the Basic Research laboratories of Toray Industries Inc., is a hydrogel polymer with poly(ethylene oxide) (PEO) chains. PEO chain lengths of 100 were graft copolymerized to polyvinylchloride (PVC) containing dithiocarbamate groups. The long-chained PEO is flexible and has a high affinity for water. It is suggested that the long-chain PEO exists on the surface of the polymer contributing significantly to the element of volume restriction and osmotic repulsion effect, which prevents the adsorption of blood components.

The SAEKA [7] is a chemically modified keratine derivative (low sulfur amino-ethyl keratin) which is composed of native polypeptides and is adsorbable *in vivo*. The polymer was supplied by Professor Dr H. Inagaki, Institute for Chemical Research, Kyoto University, 611 Gokasho, Uji, Kyoto, Japan. The SAEKA-UV polymer is a cationic keratin derivative that has been crosslinked by ultraviolet irradiation. The SAEKA-Hep-GA polymer is a heparin-ion complex polymer of the cationic SAEKA, which was crosslinked with 2.5% glutaraldehyde in an aqueous solution. The SAEKA-Hep-GA polymer contained approximately 11% heparin by weight. It is adsorbable *in vivo* and can release heparin slowly.

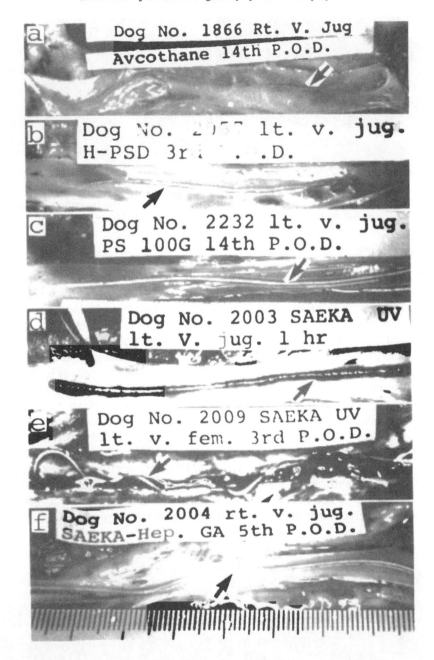

Figure 2. Macroscopic views of the specimens in the veins. Arrows indicate the polymer coated sutures. (a) Avcothane, 14 days after the insertion; no thrombus is observed on the suture. (b) H-PSD, 3 days; no thrombus is observed on the suture. Some fresh mural thrombi are on the vessel wall near the suture. (c) PS-100G, 14 days; no thrombus is observed on the suture. (d) SAEKA-UV, 1 h; the suture is covered with fresh thrombus along the entire length. (e) SAEKA-UV, 3 days; the lumen of the vein is filled with a fresh thrombus along the suture. (f) SAEKA-Hep-GA, 5 days; no thrombus is observed on the suture.

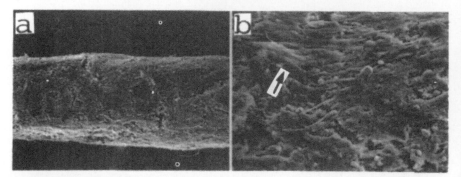

Figure 3. Scanning electron micrographs of the Avcothane coated suture surface, 7 days after the insertion. (a) The surface is covered with fibrous tissue, ×60; (b) Higher magnificational view of the surface; the fibrous tissue is covered with endothelial-like cells (arrow), ×240.

3. RESULTS

As to the avcothane, no thrombus was macroscopically observed along the entire length of the coated suture from 1 h to 14 days after the insertion (Fig. 2a). Visualized by scanning electron microscopy, many platelets and leucocytes with small amount of fibrin deposition were noticed on the surfaces of the sutures obtained from the 1 h to 5-day tests. The suture obtained after 7 days test was completely covered with a thin layer of fibrin (Fig. 3a). Occasionally, the fibrin layer was organized into a neointima with endothelial-like cells (Fig. 3b).

In the case of H-PSD, all of the specimens obtained after 1 h to 14 days showed no thrombus nor any blood components on the surface, macroscopically (Fig. 2b) and light macroscopically. Scanning electron microscopic examination of the surfaces showed numerous platelets sporadically distributed over the surfaces, but no fibrin deposition was noticed. Only serum protein adhered on the surface (Fig. 4a and b). With transmission electron microscopy, the surfaces of the sutures obtained from the 1-h to 14-day tests were covered with a thin layer of a non-specific serum protein with a thickness of 20–500 nm. A representative transmission electron micrograph is shown in Fig. 5.

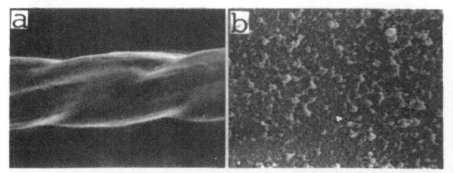

Figure 4. Scanning electron micrographs of the H-PSD coated suture surface, 1 day after the insertion. (a) No thrombus is observed on the surface at all, ×75; (b) High magnificational view of the surface. Serum protein adhered on the surface. No fibrin deposition is noticed, ×3000.

Figure 5. Transmission electron micrograph of a thin section of the H-PSD coated suture, near the surface, 14 days after the insertion. The surface is covered with serum protein. The bar indicates 1 μm, ×29000.

In the case of PS-100G, no thrombus was observed on the polymer coated sutures in all specimens by macroscopic observation (Fig. 2c). By means of scanning electron microscopy, some platelets were noticed on the surfaces of all the sutures obtained (Fig. 6a and b). With transmission electron microscopy, small amounts of serum protein were noticed on the surface of the 1 h test suture. The surfaces of the sutures obtained after 1–14 days were covered with serum protein, but the state of it was different on each surface. On the surface of the 1-day test, a thick layer of serum protein had adhered; however, the thickness

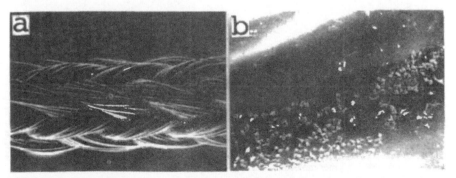

Figure 6. Scanning electron micrographs of the PS-100G coated suture surface, 7 days after the insertion. (a) No thrombus is observed on the surface, ×60. (b) Higher magnificational view of the surface; numerous platelets and leucocytes are observed on the depressions of the multifilament suture. Platelets are not aggregated. There is no fibrin deposition, ×850.

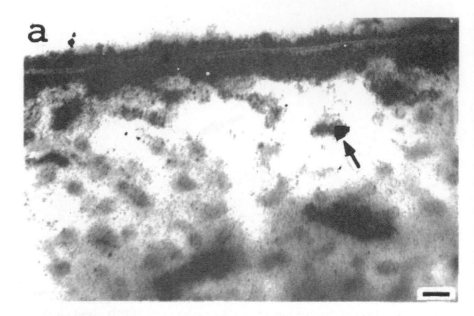

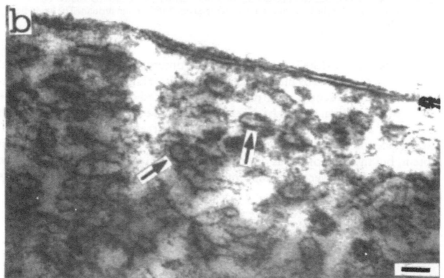

Figure 7. Transmission electron micrographs of thin sections of the PS-100G coated sutures. (a) The suture, 1 day after the insertion. Serum protein is observed on the polymer surface and inside the polymer (arrow). A translucent line is noticed at the polymer surface. The bar indicates 0.1 μm, $\times 73\,000$. (b) The suture, 14 days after the insertion. Serum protein is noticed on the surface and inside the polymer. The translucent line is clearly noticed on the surface. It forms a special double membrane like a plasma membrane of cells. Serum protein inside the polymer (arrows) shows the shapes of PS-100G polymer molecules. The bar indicates 0.1 μm $\times 81\,000$.

of it decreased gradually with the passage of time. Figure 7(a) shows the 1-h test surface. A thick layer of serum protein was noticed on the surface and a small amount of it infiltrated into the polymer from the surface. Figure 7(b) shows a high power magnification of the transmission electron micrograph of the specimen removed after 14 days. A thin layer of serum protein was observed on the surface as well as deeply infiltrated into the polymer inside. The surface of the polymer was covered with a double membrane similar to a cell plasma membrane. The serum protein inside the polymer was stained with osmium tetraoxide and lead citrate, and was observed to have invaded into the interstices of each PS-100G polymer. The shape of each molecule of the PS-100G polymer was recognized by the staining of the serum protein in the polymer.

In the case of SAEKA-UV, the coated suture after 1 h was macroscopically covered with a heavy layer of thrombus. Under light microscopy, the suture was observed to be covered with a fresh thrombus of 200–600 μm in thickness (Fig. 8). After 3 days, the blood vessel was occluded by a fresh thrombus around the suture (Fig. 2e).

In the case of SAEKA-Hep-GA, all specimens tested after 1 h to 14 days showed no thrombus macroscopically, as had been the case with Avcothane and H-PSD coated sutures (Fig. 2f). Under light microscopic observation, the SAEKA was observed as an amorphous red-colored substance following staining with eosin. In the specimen after 3 days, the polyester multifilaments of the

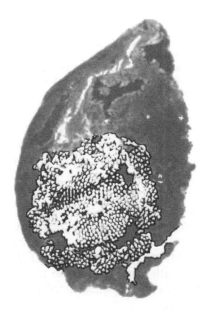

Figure 8. Photomicrograph of a cross section of the SAEKA-UV coated suture, 1 h after the insertion. The suture is surrounded with fresh, thick thrombus. Hematoxilin & Eosin stain (H & E), ×75.

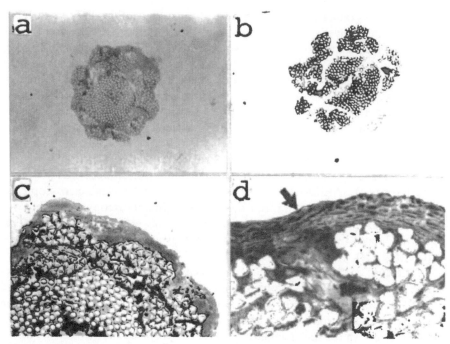

Figure 9. Photomicrographs of cross sections of the SEAKA-Hep-GA coated sutures. (a) The suture, 3 days after the insertion. Polyester multifilament suture is coated with the SAEKA-Hep-GA polymer. No thrombus is observed on the surface. H & E, ×75. (b) The suture, 5 days after the insertion. The polymer is almost absorbed from the surface. Small amount of it still remained in the interstices of the filaments of the polyester. There is no fibrin deposition around the suture. H & E, ×90. (c) The suture, 7 days after the insertion. The polymer is already absorbed. The polyester multifilaments are covered with fresh fibrin layer. Leucocytes are noticed in the fibrin layer. H & E, ×150. (d) The suture, 14 days after the insertion. The fibrin layer is changed into a neointima-like tissue. The surface is covered with a layer of endothelial-like cells (arrow). H & E, ×200.

suture were embedded in the SAEKA (Fig. 9a). In the specimen after 5 days, most of the SAEKA around the suture was already absorbed. Each polyester filament was naked and exposed to the surface. Only very small amounts of the SAEKA still remained in the interstices of the polyester multifilaments. In such a rough surfaced suture, no thrombus adhered on it (Fig. 9b). Ionically bound heparin was still releasing from the remaining SAEKA and prevented the precipitation of the fibrin around the suture. The suture obtained after the 7-day test was covered with a thin layer of fibrin. There was no SAEKA remaining in the interstices of the polyester filaments (Fig. 9c). The suture obtained after 14-day test was covered with a neointima-like tissue. The surface of it was organized and covered with a layer of endothelial-like cells (Fig. 9d). There was no thrombus on the surface.

All the spectroscopic observations are summarized in Table 1. Some of the sutures, such as the 14-day test of SAEKA-Hep-GA and Avcothane, were observed to be covered with neointima-like tissue. In these cases, the polymers no longer faced the blood stream.

Table 1.
Results of antithrombogenicity of five materials

Polymer and observation method		1 h	1 day	3 days	7 days	14 days
Avcothane	Macro	−	−	−	−	−
	LMS	−	−	+	neointima	neointima
	SEM	+	+	+ +	neointima	neointima
	TEM	+	+ +	+ + +	neointima	neointima
H-PSD	Macro	−	−	−	−	−
	LMS	−	−	−	−	−
	SEM	+	+	+	+	+
	TEM	+	+	+	+	+
PS-100G	Macro	−	−	−	−	−
	LMS	−	−	−	−	−
	SEM	+	+	+	+	+
	TEM	+	+	+	+	+
SAEKA-UV	Macro	+ +	+ +	+ + +	+ + +	+ + +
	LMS	+ + +	+ + +	+ + +	+ + +	+ + +
	SEM	+ + +	+ + +	+ + +	+ + +	+ + +
	TEM	+ + +	+ + +	+ + +	+ + +	+ + +
SAEKA-Hep-GA	Macro	−	−	−	−	−
	LMS	−	−	−	+ +	neointima
	EM	+	+	+	+ +	neointima
	TEM	+	+	+	+ + +	neointima

4. DISCUSSION

Each methods of evaluation currently used have both advantages and disadvantages. For example, the IVC ring method [8], the intra vascular magnetic suspension method [9] and the atrial 'sword' method [10] require complicated procedures such as thoracotomy of laparotomy, so that there is some fear that inexperienced researchers may lead to disastrous results. As an easier procedure, the insertion method of a small ring into the juglar vein [11] has obtained a good appraisal. It is possible to test several materials in one animal by this method. However, during its procedure, which involves vessel clamping, venotomy and insertion of the ring, the vessel wall is injured. This injury often leads to coagulation and thrombus formation in the inserted area. Besides, the test materials need to be fabricated into the special shape and size to the ring prescribed. Another easy method, the IVC indwelling catheter method [12], which does not require very much surgical skill, has also been used. Only a single material, however, can be evaluated per animal using this method. Moreover, the catheter's size and rigidity may injure the vessel wall excessively. The injured area may induce thrombus formation around the catheter.

In our method described in this paper, a pliable suture coated with the test polymer can float in the blood stream. It may touch the vessel wall or the valves; however, the mechanical stress of the suture rarely causes injury to the vessel wall. Should injury happen, collagen fibrils or elastica beneath the endothelial cells of the wall will face the blood stream, a small thrombus will be formed just on the injured areas but not on the suture surface. Figure 2(b) showed the mural thrombi near the suture.

One of the characteristics of this method was that the peripheral venous system was adopted as a site of the interaction between the blood and polymers. The test polymer can float in the peripheral venous blood. In general, the blood flow of the peripheral vein is quite slow. Once a thrombus forms inside it, it will grow into a larger thrombus, which will occupy the whole lumen. Thus, it will often lead to an occlusion. Therefore, the evaluation test in a venous system is more restrictive as compared to that in an arterial system. In the arterial system, a small mural thrombus seldom leads to arterial occlusion, because a mural thrombus cannot grow effectively in the arterial system. Many of them are carried off by the fast arterial blood flow. As shown in the cases of PS-100G and Avcothane, some leucocytes, platelets and fibrin deposition adhered on the polymer surfaces, but the venous blood flow did not wash them off. They kept attaching onto the polymer surface and developed naturally into a large thrombus. Even if the thrombus were detached from the polymer surface into the venous blood stream, it will be carried to the lung, since peripheral venous blood flows into the pulmonal artery through the inferior of superior vena cava and the right ventricle of the heart. The results of the necropsy of the animals tested indicated that no embolus nor thrombus was in the lungs, kidneys or spleens. Even in the case of an occlusion of the vessel with fresh thrombus, thrombus was only found at the site around the suture. No significant sign of thrombus wash-off was noticed in any of the test animals. Therefore, it is considered that the thrombus formed on the polymer in the peripheral venous blood retains the original situation and the relationship with the polymer. From these observations, the presence of any thrombus or blood components adhered on the polymer surface is the actual results of the relationship between the polymer and the blood.

In the case of Avcothane, a thin layer of fibrin with endothelial-like cells was observed on the 7-day test suture, indicating that the antithrombogenicity of Avcothane is relatively weak and is effective only within 1 week. In the case of H-PSD, a simple relationship with the blood was noticed, that is a thin layer of serum protein led to the stable interaction.

In the case of PS-100G, the infiltration of serum protein into the intersticial areas of the polymer was observed with the passage of time. The double membrane of the polymer surface has not been analyzed for its composition, the origin nor the role of the antithrombogenic property of the polymer. However, these observations should give us a better understanding of the real interaction of the serum protein and the polymer. We are presently investigating the membrane morphologically, biochemically and functionally.

In the case of the SAEKA-Hep-GA polymer, we could observe the sequence of the antithrombogenic activity, i.e. the prevention of the fibrin deposition by the antithrombogenicity due to the slow release of heparin, the dissolving the polymer, the fibrin deposition after the dissolving the polymer and the endothelialization on the fibrin layer. Thus, the suture got permanent antithrombogenicity by the sequential combination of the artificial antithrombogenicity by the slow release of heparin and the natural one of the endothelialization. This induction of the natural antithrombogenicity can be a very useful method for implantable artificial organs like a vascular graft. From these observations, it is suggested that the characteristic advantage of our method is that we can observe the interaction between the host and test polymer at physiological conditions over a long period.

This method proved to be very valuable for analyzing the relationship between polymers and the blood, as it brought forth new evidence. In particular, it has many advantages as a screening method: (i) The procedure is simple and easy compared with any other methods reported. (ii) Surgical intervention affecting the animal is minimal. (iii) Injury to the blood vessel is negligible. (iv) Four polymers can be tested at the same time. (v) It is unnecessary to put the test polymer into a special shape. (vi) The test is strict. (vii) Specimens for light and electron microscopy can be obtained from one experiment.

5. CONCLUSION

A newly developed *in vivo* method for the estimation of blood compatibility was demonstrated. The role of *in vivo* methods is to observe the real, long-term relationship between the polymer and the host. All the phenomena such as the adaptation activity to the polymer, the healing process against the invasion of the polymer and the foreign body reaction will be easily observed by this method.

REFERENCES

1. Noishiki, Y., An *in vivo* test for evaluation of blood compatibility by insertion of the polymer coated suture into the peripheral vein. *Jpn. J. Artif. Organs* 11, 974–977 (1982).
2. Noishiki, Y., A new *in vivo* evaluation method of antithrombogenicity using a polymer-coated suture inserted into the peripheral vein. *J. Bioact. Comp. Poly.* 1, 147–161 (1986).
3. Nylas, E. and Ward, R. S. Jr., Development of blood compatible elastomers. V. Surface structure and blood compatibility of Avcothane elastomers. *J. Biomed. Mater. Res.* 11, 69–84 (1973).
4. Tanzawa, H., Mori, Y., Harumiya, N., Miyama, H., Hori, M., Ohshima, N. and Idezuki, Y., Preparation and evaluation of a new antithrombogenic heparinized hydrophylic polymer for using cardiovascular system. *Trans. Am. Soc. Artif. Intern. Organs* 19, 188–194 (1973).
5. Miyama, H., Harumiya, N., Mori, Y. and Tanzawa, H., A new antithrombogenic heparinized polymer. *J. Biomed. Mater. Res.* 11, 251–265 (1977).
6. Mori, Y., Nagaoka, S., Takiuchi, H., Kikuchi, T., Noguchi, N., Tnazawa, H. and Noishiki, Y., A new antithrombogenic material with long polyethyleneoxide chains. *Trans. Am. Soc. Artif. Intern. Organs* 28, 459–463 (1982).
7. Ito, H., Miyamoto, T., Noishiki, Y. and Inagaki, Y., High-molecular fraction of low sulfur S-carboxymethyl keratin. *Seni-Gakkaishi* 34 , 157–165 (1978).
8. Gott, V. L., Ameri, M. L., Whiffen, J. D., Leinger, R. I. and Falb, R. D.: Newer thrombo-resistant surfaces with a new *in vivo* technique, *Surg. Clin. N. Am.* 47, 1443–1452 (1967).
9. Lederman, D. M., Cumming, R. D., Petschek, H. E., Chiu, T. H., Nylas, E., Zalzman, E., Collins, R. E. C. and Coe, N. P., The intravascular magnetic suspension of a test device for *in vivo* hemocompatibility. *Trans. Am. Soc. Artif. Intern. Organs* 22, 545–553 (1976).
10. Jacob, L. A., Klopp, E. and Gott, V. L., Study on the fibrinolytic removal of thrombus from prosthetic surfaces. *Trans. Am. Soc. Artif. Intern. Organs* 14, 63–68 (1968).
11. Whalen, R. L., Jeffery, D. L. and Norman, J. C., A new method of *in vivo* screening of thromboresistant biomaterials utilizing flow measurement. *Trans. Am. Soc. Artif. Intern. Organs* 19, 19–24 (1973).
12. Idezuki, Y., Watanabe, H., Hagiwara, M., Kanasugi, K., Mori, Y., Nagaoka, S., Hagino, M., Yamamoto, K. and Tanzawa, H., Mechanism of antithrombogenicity of a new heparinized hydrophilic polymer: Chronic *in vivo* studies and clinical application. *Trans. Am. Soc. Artif. Intern. Organs* 21, 436–449 (1975).

Multiphase Biomedical Materials, pp. 73–87 (1989)
T. Tsuruta and A. Nakajima (Eds)
1989 VSP.

Chapter 5

Hybrid biomaterials incorporated with living cells in modified collagens

TOSHIHIRO AKAIKE,[1] SHUNJI KASAI,[1] SHINJI NISHIZAWA,[1]
AKIRA KOBAYASHI[1] and TERUO MIYATA[2]
[1]*Department of Material Systems Engineering, Faculty of Technology, Tokyo
University of Agriculture and Technology, Nakamachi, Koganei, Tokyo 184, Japan*
[2]*Japan Biomedical Material Research Center, Nakane, Meguro, Tokyo 152, Japan*

Summary—In order to design hybrid materials such as hybrid organs and biosimulators where living cells are incorporated in a collageneous matrix, the interactions of cells [e.g. fibroblasts (L-cells), macrophages and hepatocytes] with chemically or physically modified collagen, e.g. methylated, succinylated and heat-denatured collagens, were examined. It was found that there is a significant difference of cellular response to the modification of collagen among fibroblasts (L-cell), macrophages and hepatocytes. Recognition mechanisms mediated by collagen receptors on cellular surfaces are discussed relating to the high affinity of hepatocytes to unmodified collagen.

1. INTRODUCTION

Collagen is a major component of the extracellular matrix in many tissues, with which most cells, fibroblasts or epithelial cells are known to have interactions. Experiments *in vitro* have clearly demonstrated that collagen can influence proliferation, differentiation and migration. However, the molecular mechanisms of how collagen may communicate with cells have not been clarified so far.

Recently the studies on artificial organs have been developed, together with studies on artificial organs incorporated with cells peculiar to each organ, that is to say, the studies on the hybrid organs have been actively conducted. For instance, the development of a hybrid organ for the liver in which hepatocytes are incorporated in keeping cell metabolic activities has been attempted. In this case, it is important to attach the cells to the substrata without losing the *in vivo* cell activity and the success depends on what kind of the substratum is used. As the substratum of liver cells, collagen extracted from the liver is reported to be excellent [1].

So-called methods of cell technology such as cell culture and cell isolation occupy one of the important scientific technical fields along the advancement of life science. That is to say, effective cell culture and the isolation techniques for cells are the most important techniques in cell technology for the production of biologically active substances derived from cells as well as *in vitro* analysis of cellular functions. In the culture of animal cells, it is important to select the substratum with a good adhesive property for cells in order to keep the original cell activity or in order to make the cells proliferate. The addition of fetal calf serum (FCS) to the culture medium is usually required in order to maintain the cell activity or to proliferate cells. However, recently, owing to the difficulties of the availability and expense of FCS, the establishment of serum free cell culture methods have been earnestly desired.

It is an object of the present work to estimate the cell–material interactions in order to provide collagen substratum to which animal cells can adhere effectively in a system in the presence or absence of FCS. Therefore, in this study a variety of chemically or physically modified collagens were prepared, and the attachment behavior of some cell species to them was evaluated to clarify the role of the chemical and physical structure of collagen matrix. Fibroblasts (L-cell), macrophages (mouse peritoneal exudate cells) and hepatocytes (rat liver parenchymal cells) were used.

2. MATERIALS AND METHODS

2.1. Preparation of collagen and chemical modification of collagen side chains
Calf-skin insoluble collagen was solubilized by pepsin at pH 3 at room temperature. The solubilized collagen was purified by filtration through a 1-μm Millipore filter and by precipitation at pH 7. Monomeric dispersed collagen was chemically modified by methylation or succinylation [2]. A collagen solution was prepared according to the method of Michalopoulos and Peitot [3]. Unmodified succinylated or methylated collagen (0.1 g) was solubilized in 50 ml of sterile 0.1% acetic acid in H_2O and stirred at 4°C for 48 h. The mixture was centrifuged at 10 000 g for 30 min and the supernatant was used as the stock collagen solution.

2.2. Collagen-coated substrata
Collagen-coated substrata were prepared according to the method of Grinnell and Minter [4], with slight modifications. Aliquots (0.8 ml) of a collagen solution, 0.5 mg/ml in 0.1% acetic acid, were incubated in 35-mm Nunc culture dishes for 20 min at 22°C, after which the dishes were rinsed three times with Hanks' balanced salt solution (HBSS).

2.3. Heat-denatured collagen-coated substrata
Aliquots (0.8 ml) of a collagen solution, 0.5 mg/ml in 0.1% acetic acid, were heated to 50°C for 20 min. The solutions were then cooled to 37°C and incubated in dishes for 20 min at 22°C, after which the dishes were rinsed three times with HBSS. In some experiments, plasma fibronectin dissolved in HBSS (100 μg/ml) were added in 0.5-ml portions to non-heat-denatured or heat-denatured collagen-coated substrata, incubated overnight at 4°C, and washed twice with HBSS before addition of the cells.

2.4. Purification of plasma fibronectin

Plasma fibronectin was purified from human plasma by gelatin affinity chromatography [5]. Fifty milliliters of human plasma was passed through a 70 cm^2 column of plain Sepharose 4B before fractionation on gelatin–Sepharose 4B to remove any material binding to Sepharose. The same volume of effluent plasma from plain Sepharose 4B was passed through a 1.6×20 cm^2 column of gelatin–Sepharose 4B in phosphate-buffered saline (PBS) (pH 7.2), containing 0.01 M sodium citrate at room temperature. After the column was washed with PBS containing 0.01 M sodium citrate and 1 M urea in 0.05 M Tris–HCl (pH 7.5), the adsorbed fibronectin was eluted with 4 M urea in 0.05 M Tris–HCl (pH 7.5). The eluate was dialyzed at room temperature against PBS and stored at $-20°C$. The isolated plasma fibronectin was 90% pure, as estimated by sodium dodecyl sulfate–polyacrylamide gel electrophoresis.

2.5. Preparations of cells and the kinetics of cell attachment

2.5.1. Cells. L-cells were grown in stationary culture in Eagle's MEM supplemented with 10% FCS and 60 μg of kanamycin per ml. L-cells were dispersed from the monolayer with 0.25% trypsin and 0.02% ethylenediaminetetraacetate (EDTA) in PBS. The detached cells were harvested by centrifugation, washed once with HBSS containing BSA at 10 mg/ml, and resuspended in Eagle's MEM with or without 10% FCS. L-cells (5×10^5 cells) in 1.5 ml of adhesion medium were placed in various collagen-coated dishes. The dishes were incubated at 37°C in a 5% CO_2 atmosphere. The number of adherent cells was determined at various times by counting the number of non-adherent cells recovered in the medium and subtracting this number from the number of cells initially placed in the dish.

2.5.2. Mouse peritoneal exudate cells (peritoneal macrophages). Female CD-1 mice, 6–10 weeks of age (from Charles River Japan, Inc.) were used. Peritoneal exudate cells (PECs) were harvested by massage of the peritoneal cavity after injection of 5 ml of cold HBSS and aspiration of the exudate using a syringe with a 25-gauge needle. PECs were centrifuged, suspended in RPMI 1640 medium (Nissui Seiyaku Co.) either with or without 10% fetal bovine serum (Grand Island Biological Co.) and kept on ice until use.

PECs (containing 1×10^6 macrophages) in 1.5 ml of adhesion medium were placed in various collagen-coated dishes. The dishes were incubated at 37°C in a 50% CO_2 atmosphere. The number of adherent cells was determined at various times by counting the number of non-adherent cells recovered in the medium.

Macrophages and lymphocytes were distinguished by standard morphologic criteria by phase-contrast microscopy, or by their characteristic morphologic appearance in Giemsa- and non-specific esterase-stained cytocentrifuge preparations.

2.5.3. Rat hepatocytes (rat liver parenchymal cells). Rat hepatocytes were isolated from adult female Sprague–Dawley rats (150–250 g) by *in situ* perfusion

of the liver with collagenase (Type I) according to Seglen [6]. Isolated hepatocytes were suspended in Williams medium E with 10% fetal bovine serum.

In each experiment of assay of cell adhesion, the cells were seeded in various
collagen-coated dishes with or without 10% FCS. The dishes were incubated at
37°C in a 5% CO_2 atmosphere. The time-course of cell adhesion was monitored
by counting the number of non-adherent cells recovered in the medium and subtracting this number from the number of cells initially seeded in the dish. The
viability of the cells was also estimated by the number of attached cells after
incubation. The activity of tyrosine aminotransferase (TAT) of attached
hepatocytes, which is a typical liver enzyme, was assayed after 48 h according
to the method of Granner.

3. RESULTS

3.1. Net charge on unmodified and chemically modified collagen
To check for the helical stability of collagen during various chemical modification procedures, each collagen solution was assayed by circular dichroism (CD).
The CD spectrum of succinylated or methylated collagen was similar to that of
unmodified collagen, having a positive peak at 220 nm and a negative deflection
below 210 nm (Fig. 1). It was thought, therefore, that none of the chemical
modifications caused any conformational change in collagen.

Carboxyl and ε-amino groups of unmodified and chemically modified collagen were determined by hydrogen ion titration [7] and the 2,4,6-trinitrobenzene

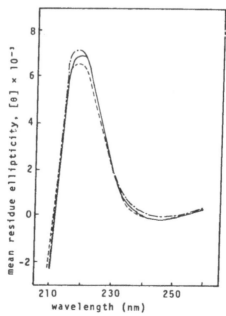

Figure 1. CD spectra of chemically modified collagen. Each collagen was dissolved in 10 mM
phosphate buffer (pH 3). Values for $[\theta]$ are deg/cm^{-2}/dmol. Unmodified collagen (——);
succinylated collagen (—·—); methylated collagen (- - -).

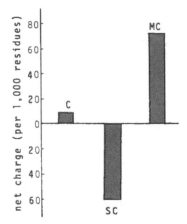

Figure 2. The net charge on unmodified and chemically modified collagen. The net charge was calculated for pH 7.3 on the basis of 1000 amino acid residues. Carboxyl and ε-amino groups of unmodified and chemically modified collagen were determined by hydrogen ion titration and the 2,4,6-trinitrobenzene sulfonic acid colorimetric method, respectively. C, unmodified collagen; SC, succinylated collagen; MC, methylated collagen.

sulfonic acid colorimetric method [8], respectively. Succinylation blocked 97% of the ε-amino groups on unmodified collagen, and this exactly corresponds to the increase in carboxyl groups [2]. Methylation, on the other hand, masked 85% of the free carboxyl groups on unmodified collagen [2]. Figure 2 shows the results of analysis of the carboxyl and ε-amino groups of unmodified and chemically modified collagens. At pH 7.3, the net charge on unmodified, succinylated and methylated collagen was approximately neutral, negative and positive, respectively.

3.2. Relationship between L-cell attachment and the molecular structure of collagens

3.2.1. Effect of chemical modifications of collagen on L-cell attachment. L-cells attached more rapidly to methylated (with a net positive charge) collagen-coated substrata than to succinylated (with a net negative charge) collagen-coated substrata in the presence or absence of serum (Fig. 3). However, as the incubation time increased, the rate of cell attachment to succinylated collagen-coated substrata was markedly increased. After 1 h of incubation more than 85% of the cells attached to succinylated collagen-coated substrata, just as in the case of methylated collagen-coated substrata in the presence of serum. In contrast, fewer cells attached to unmodified collagen, which is approximately neutral in net charge, and the attachment did not exceed 35% after 1 h of incubation.

On unmodified collagen-coated substrata, partial or no cell spreading was observed with and without serum. On the other hand, succinylation or methylation of collagen promoted cell spreading on the substrata regardless of the presence or absence of serum. These results indicate that the alteration of the net

T. Akaike et al.

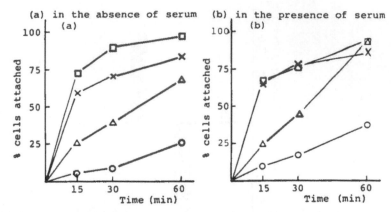

Figure 3. Time course of L-cell attachment to chemically modified collagen substrata. ○, AC; △, SAC; □, MAC; ×, non-coating.

charge of collagen markedly affects the rate of cell attachment and spreading, independently of the presence of serum.

3.2.2. Effect of the conformation of collagen on L-cell attachment. The rate of L-cell attachment to unmodified or chemically modified collagen-coated dishes was nearly the same in the presence or absence of serum (Fig. 3). In contrast, heat denaturation of collagen enhanced the rate of cell attachment in the presence of serum (Fig. 4). In serum-free medium, however, no difference of L-cell attachment was observed between native collagen and heat-denatured collagen. The results indicate that the rate of L-cell attachment to a heat-denatured collagen-coated substratum is significantly affected by serum but not

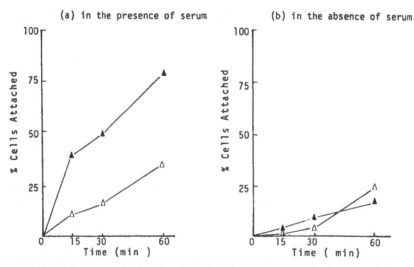

Figure 4. Attachment of L-cells to collagen-coated dishes: effect of denaturation of atelocollagen. Atelocollagen, △; heat-denatured atelocollagen, ▲.

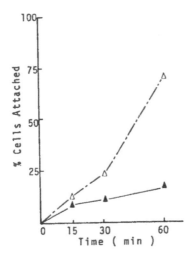

Figure 5. Attachment of L-cells to heat-denatured atelocollagen-coated dishes untreated or previously treated with fibronectin. Heat-denatured atelocollagen- and fibronectin-coated dishes, △; heat-denatured atelocollagen-coated dishes, ▲.

to a non-heat-denatured collagen-coated substratum (Fig. 4a and b). In other words, only under serum-containing conditions did there exist a significant effect of the conformational change from native collagen to the disordered form on cell attachment.

3.2.3. Effect of fibronectin in serum on L-cell attachment to heat-denatured collagen. As stated above, L-cell attachment to heat-denatured collagen was mediated by serum. A similar result was obtained for L-cell attachment to heat-denatured collagen which had been treated with purified plasma fibronectin (Fig. 5). This demonstrates the possibility that fibronectin in serum is a mediator for L-cell attachment to heat-denatured collagen.

3.3. Relationship between macrophage adhesion and the molecular structure of collagen

3.3.1. Effect of chemical modification on macrophage adhesion. When unseparated mouse PECs, containing approximately 50% macrophages and 50% lymphocytes, were incubated in chemically modified collagen-coated dishes, only macrophages adhered. Mouse peritoneal macrophages adhered rapidly to succinylated (with a net negative charge) or methylated (with a net positive charge) collagen-coated dishes (Fig. 6). In contrast, few macrophages adhered to unmodified collagen, which is approximately neutral in net charge. These results demonstrate that side-chain modifications of collagen promote the adhesion of macrophages to the substratum. The rate of macrophage adhesion to unmodified or chemically modified collagen-coated dishes was the same in the presence or absence of serum (Fig. 6a and b).

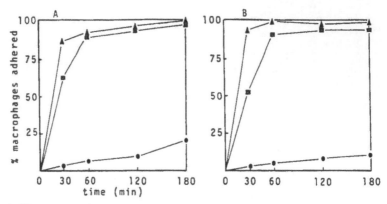

Figure 6. Time course of macrophage adhesion to chemically modified collagen substrata in the presence (A) or absence (B) of serum. ●, unmodified collagen substratum; ■, succinylated collagen substratum; ▲, methylated collagen substratum.

3.3.2. Effect of the conformation of collagen on macrophage adhesion.

As in L-cell attachment, heat denaturation of collagen also enhanced the rate of macrophage adhesion in the presence of serum (Fig. 7). The results indicate that the rate of macrophage adhesion to a heat-denatured collagen-coated substratum is significantly affected by serum, but not to a non-heat-denatured collagen-coated substratum.

In this case, the mediation of some serum factors was suggested to be the enhancement effect of cell adhesion to denatured collagen. So, in order to clarify the mechanism, macrophage adhesion to heat-denatured collagen, which had been treated with purified plasma fibronectin, was examined. Nearly the same results as obtained in L-cell attachment demonstrate the possibility that fibronectin in serum is also a mediator for macrophage adhesion to heat-denatured collagen.

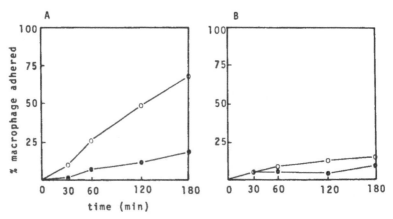

Figure 7. Time course of macrophage adhesion to non-heat-denatured (●) or heat-denatured (O) unmodified collagen substratum in the presence (A) or absence (B) of serum.

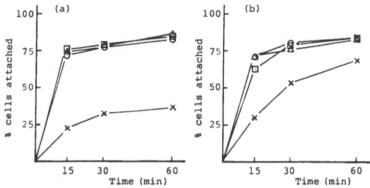

Figure 8. Time course of hepatocytes attachment to collagen substrata in the absence (a) or presence (b) of serum. O, unmodified collagen; △, succinylated collagen; □, methylated collagen substrata; ×, non-coated dish.

3.4. Relationship between hepatocyte attachment and the molecular structure of collagens

3.4.1. Effect of chemical modifications of collagen on hepatocyte adhesion.
Rat hepatocytes were incubated in chemically modified collagen-coated dishes and non-coated dishes in the presence or absence of serum. Figure 8 shows the relationship between the net charge of collagen and hepatocyte adhesion in the presence (Fig. 8a) or the absence (Fig. 8b) of 10% fetal serum. No difference

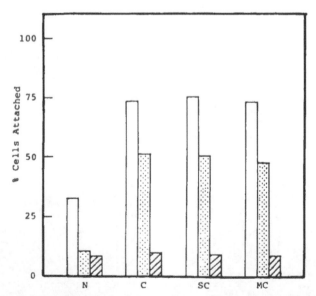

Figure 9. Effect of the divalent cations on the attachment of hepatocytes to collagenous substrata. The cell attachment determined in the absence of serum after 60 min as described in Materials and Methods. The medium contained; □, no addition; ▨, 3 mM EDTA; ▨, 2 mM EGTA. N, non-coated dish; C, unmodified collagen, SC, succinylated collagen; MC, methylated collagen.

in the cell attachment rate was observed among various collagen surfaces, while fewer hepatocytes attached to the non-coated surfaces. It was considered that hepatocytes have a higher affinity to collagen and the cell attachment is not affected by the change of net charge of collagen. It was found that hepatocyte attachment was inhibited by the addition of chelating reagents such as EDTA or EGTA to the medium (Fig. 9). It was suggested that the divalent cations, e.g. Ca^{2+} ions, stimulate the attachment of hepatocytes to substrata. The Mg^{2+} ion was thought to be more effective to hepatocyte adhesion than the Ca^{2+} ion.

On the other hand, as with the spreading activity of adherent hepatocytes, a considerable difference was observed among various kinds of collagens. The cell spreading was estimated in a phase contrast microscope. Hepatocytes were found to spread when attached to unmodified collagen and succinylated collagen independently of serum (Fig. 10B and C). Little spreading of cells was observed on methylated collagen (Fig. 10C), and the spreading of hepatocytes attached to a non-coated culture dish was found to be very low (Fig. 10A). It was interesting that little cell spreading was observed on methylated collagen in spite of retaining a high rate of cell attachment.

An important problem in the development of a hybrid artificial liver is to maintain the activities of the cells immobilized on the matrix. The activity of tyrosine aminotransferase, a typical liver enzyme (Scheme 1, TAT) after a 48-h incubation of cultured hepatocytes was also evaluated as well as the activity of glycogenolysis under serum-containing condition (Table 1).

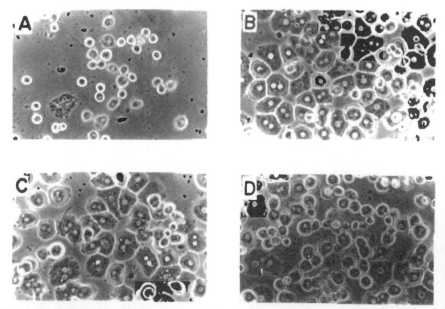

Figure 10. Morphology of hepatocytes spreading on the substrata coated with unmodified and chemically modified collagen. Cells were cultured for 7 h in the absence of serum on non-coated dish (A), unmodified collagen (B), succinylated collagen (C), methylated collagen (D).

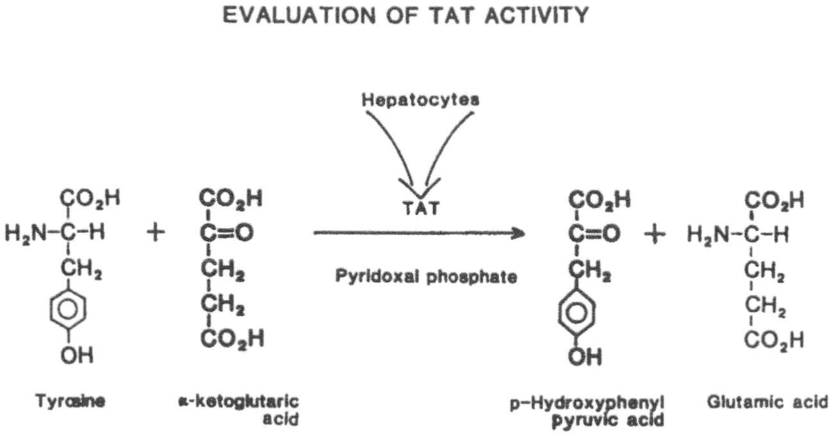

EVALUATION OF TAT ACTIVITY

Hepatocytes

$$\underset{\text{Tyrosine}}{\underset{\overset{|}{\underset{\overset{|}{\text{OH}}}{\bigcirc}}}{\underset{\overset{|}{\text{CH}_2}}{\overset{\text{CO}_2\text{H}}{\underset{|}{\text{H}_2\text{N}-\text{C}-\text{H}}}}}} + \underset{\substack{\alpha\text{-ketoglutaric}\\\text{acid}}}{\overset{\text{CO}_2\text{H}}{\underset{\overset{|}{\text{CO}_2\text{H}}}{\underset{\overset{|}{\text{CH}_2}}{\underset{\overset{|}{\text{CH}_2}}{\overset{|}{\text{C}=\text{O}}}}}}} \xrightarrow[\text{Pyridoxal phosphate}]{\text{TAT}} \underset{\substack{p-\text{Hydroxyphenyl}\\\text{pyruvic acid}}}{\underset{\overset{|}{\underset{\overset{|}{\text{OH}}}{\bigcirc}}}{\underset{\overset{|}{\text{CH}_2}}{\overset{\text{CO}_2\text{H}}{\overset{|}{\text{C}=\text{O}}}}}} + \underset{\text{Glutamic acid}}{\underset{\overset{|}{\text{CO}_2\text{H}}}{\underset{\overset{|}{\text{CH}_2}}{\underset{\overset{|}{\text{CH}_2}}{\overset{\text{CO}_2\text{H}}{\underset{|}{\text{H}_2\text{N}-\text{C}-\text{H}}}}}}}$$

Scheme 1.

Table 1.
TAT and glycogenolysis activities of hepatocytes cultured on several substrata

Substratum	TAT induced[a] (%)	Glucose released (μmol/h/mg protein)
non-coated dish	100	0.59
unmodified collagen	157	0.55
succinylated collagen	178	0.74
methylated collagen	141	0.49

[a] TAT activity expressed as a percentage for the activity of non-coated dish. Each volume represents the average of five experiments. The induction of TAT on non-coated dish was 10–20 mU/mg protein.

It was observed that the induction of TAT in cultured hepatocytes was different on various substrata. The induction of TAT was generally higher on collagenous substrata than non-coated dishes.

The highest induction of TAT was observed when succinylated collagen was used as substrata. Moreover, succinylated collagen substrata also enhanced the release of glucose. Activities of hepatocytes are affected by the addition of hormones, e.g. TAT activity was enhanced by dexamethasone and glycogenolysis was stimulated by glucagon. In this study, it was indicated that activities of hepatocytes were affected by the difference of substrata without hormones, although this effect was not as large when compared to hormones.

Maintenance of hepatocytes during culture on various substrata was examined under serum-free condition (Fig. 11). Cell survival can be measured conveniently by determining the number of attached cells, because only viable cells remain attached to the substrata. With hepatocytes cultured on noncoated dish, about 70% of cells detached after 3 days in the absence of serum. On the other hand more than 50% of cells were retained on various collagens. In particular about 75% of cells attached on AC were retained after 3 days incubation.

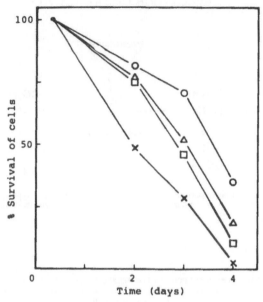

Figure 11. Maintenance of hepatocytes cultured on several surfaces in the absence of serum. Attached cells cultured for 7 h was taken as 100%. O, AC; △, SAC; □, MAC; ×, non-coated.

For the maintenance of hepatocytes, various serum factors had been thought to be very important. The results indicate that the substratum for the culture of cells is another important factor.

3.4.2. Effect of conformation of collagen on hepatocyte attachment. A remarkable decrease in cell attachment was observed on heat-denatured collagen-coated dishes as compared with native collagen-coated dishes (Fig. 12). Not only spreading but also the viability of hepatocytes on denatured collagen was also decreased (Fig. 13). The results indicate that the helical structure of collagen is one of the very important factors for high attachment, long maintenance and the spreading of hepatocytes.

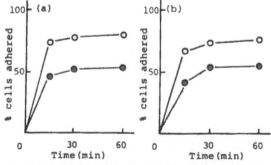

Figure 12. Effect of heat denaturation of collagen on hepatocytes attachment in the absence (a) or presence (b) of serum. O, Native collagen; ●, heat-denatured collagen.

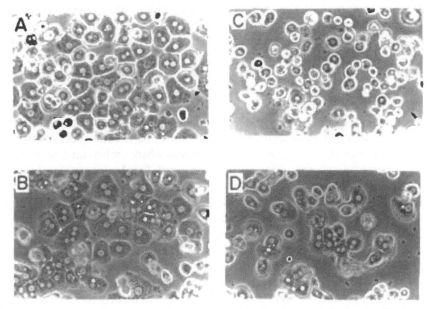

Figure 13. Effect of the conformation of collagen on hepatocytes attachment. Hepatocytes spreading on the substrata coated with native collagen (A and B) or heat-denatured collagen (C and D) in the absence (A and C) or presence (B and D) of serum. Cells were cultured for 7 h

4. DISCUSSION

In this chapter it is suggested that the adhesion of fibroblasts (L-cell) and macrophages to collagen is markedly affected by both structural alterations of collagen and the presence or absence serum.

Chemical modification of the collagen side chains was used to investigate the importance of the net charge of collagen in the adhesion of fibroblasts and macrophages. Enzyme-solubilized collagen, with a rigid triple-helical structure, has ε-NH$_2$ and COOH groups available for chemical modification for special applications to biological materials. For example, succinylated and methylated collagens become antithrombogenic and thrombogenic, respectively, compared with unmodified collagen. At pH 7.3, the net charge on unmodified collagen is approximately neutral (Fig. 2). Succinic anhydride can react with collagen, replacing almost 97% of the amino groups by carboxyl groups. Of the free carboxyl groups of collagen, 85% can be esterified with anhydrous methanol containing HCl. Thus, collagen can be converted into a highly charged protein, either negative or positive (Fig. 2), without a change in the basic triple-helical structure of the molecule (Fig. 1). As shown in Figs. 3 and 6, few fibroblasts and macrophages adhered to unmodified collagen-coated substrata, but chemical modification of the collagen side chains resulted in an enhancement of the adhesion of both cell species to succinylated collagen-coated substrata as well as methylated ones. It is very interesting that succinylation and methylation of collagen induce the enhancement of two different cells adhesion. Since cells can adhere and grow on both glass and tissue culture dishes (negatively charged) and polylysine coated surfaces or cytodex 1 microcarriers (Pharmacia Fine Chemicals Co.) (positively charged), the basic factor governing adhesion and growth of cells might be suggested to be the density of the charges on the culture surface rather than the polarity of the charges. Adhesion of L-cells and macrophages to the substratum coated with native collagen occurs to the same extent both in the presence and absence of serum (Figs. 3 and 6), while adhesion to the substratum coated with heat-denatured collagen is significantly enhanced in the presence of serum (Figs. 4 and 7). A serum factor called fibronectin has been shown to promote cell attachment and spreading when applied on a substratum [4, 9]. Therefore, whether fibronectin is required for the interaction between L-cells or macrophages and the substrata coated with denatured collagen was investigated. As shown in Fig. 5, fibronectin bound to denatured collagen substrata mediates adhesion of L-cells and macrophages to the substrata.

On the other hand, the attachment behavior of rat hepatocytes was quite different from that of L-cells and macrophages. The remarkable characteristics in hepatocyte attachment are high affinity to unmodified collagen, no effect of change in the net charge of collagen and a considerable decrease in cell attachment to heat-denatured collagen. The behavior of cell attachment to chemically and physically modified collagens in this study are summarized in Table 2. It has been known that rat hepatocytes have specific affinity sites to native collagen, so-called collagen receptors, and that rapid attachment and spreading of hepatocytes to and on a collagen matrix is cooperatively mediated by the collagen receptors. It was suggested that collagen receptor cannot sensitively recognize the change in net charge of collagen by chemical modification but can identify the conformational change of collagen by heating.

On the other hand, L-cells and macrophages are considered to have no or very few collagen receptors and to have a higher affinity to denatured collagen mediated only by the interaction between fibronectin and its receptor rather than to native collagen. Rapid adhesion of L-cells and macrophages to succinylated or methylated collagen might be explained on the basis of a non-specific interaction, e.g. an electrostatic one, other than collagen–receptor mediation.

The expectation for hybrid biomedical devices such as hybrid organs and

Table 2.
Cellular attachment to modified collagen

Cell	AC	(2+)	SAC	(2+)	MAC	(2+)	Heat denaturation
platelet	(+)[a] −		(−−)[a] −−		(++)[a] ++		⟶
macrophage	−	(O)	+	(O)	++	(×)	⟶ (serum or FN)
L-cell	−	(O)	+	(O)	++	(×)	⟶ (serum or FN)
HeLa	+	(O)	++	(O)	++	(×)	⟶
hepatocyte	++	(O)	++	(O)	++	(O)	⟶

[a] Ref. 10.

hybrid biosensors is growing higher and higher because the complete replacement of biological functions (cells and organs) by synthetic materials is very difficult. In recent years much interest has been shown in the role of biological fiber structure, e.g. collagen and fibrin, on cellular function. Collagen is a major component of the extra-cellular matrix in living tissue and has been a very promising biomaterial.

In this research, it was confirmed that we can control the attachment and spreading of various types of cells by chemical or physical modification of collagen substrata. Therefore, modified collagens are expected to offer us expanded applications as hybrid biomaterials incorporated with living cell.

REFERENCES

1. Rojkind, M., *et al.*, Connective tissue biomatrix: Its isolation for long-term culture of normal rat hepatocytes. *J. Cell. Biol.* **87**, 255 (1980).
2. Wang, C. L., Miyata, T., Weksler, B., Rubin, A. S. and Stenzel, K. H., Collagen induced platelet aggregation and release. I. Effect of side-chain modifications and role of arginyl residues. *Biochem. Biophys. Acta.* **544**, 555–567 (1978).
3. Michalopoulos, G. and Pitot, H. C., Primary culture of parenchymal liver cells on collagen membrane. Morphological and biochemical observations. *Exp. Cell. Res.* **94**, 70–78 (1975).
4. Grinnell, F. and Minter, D., Attachment and spreading of baby hamster kidney cells to collagen substrata: Effects of cold-insoluble globulin. *Proc. Natl. Acad. Sci. USA* **75**, 4408–4412 (1978).
5. Engvall, E., Ruoslahti, E. and Miller, E. J., Affinity of fibronectin to collagens of different genetic types and to fibrinogen. *J. Exp. med.* **147**, 1584–1595 (1978).
6. Seglen, P. O., Preparation of isolated rat liver cells. *Methods Cell Biol.* **13**, 29–83 (1976).
7. Tanford, C., The interpretation of hydrogen ion titration curves of protein. *Adv. Protein Chem.* **17**, 69–165 (1962).
8. Kakade, M. L. and Liener, I. E., Determination of available lysine in proteins. *Anal. Biochem.* **27**, 273–280 (1969).
9. Klebe, R. J., Isolation of a collagen-dependent cell attachment factor. *Nature* **250**, 248–251 (1974).
10. Miyata, T., Schwartz, A., Wang, C. L., Rubin, A. L. and Stenzel, K. H., Deposition of platelets and fibrin on chemically modified collagen hollow fibers. *Trans. Am. Soc. Artif. Int Organs* **22**, 261 (1976).

Multiphase Biomedical Materials, pp. 89–104 (1989)
T. Tsuruta and A. Nakajima (Eds)
© 1989 VSP.

Chapter 6

Hybridization of natural tissues containing collagen with biocompatible materials: adhesion to tooth substrates

NOBUO NAKABAYASHI

Institute for Medical and Dental Engineering, Tokyo Medical and Dental University, Kanda, Tokyo 101, Japan

Summary—The connection of artificial materials with natural tissues is one of the most important matters in biomaterials and artificial organ research. Adhesion to tooth substrates must be carried out in dental treatment, but so far this has been impossible. When such adhesive materials to the tissues are prepared, dental treatment will be changed considerably.

Dental adhesives were developed by biocompatible methacrylates which had both hydrophobic and hydrophilic groups. They were HNPM, *p*-substituted Phenyl-P and 4-META. These methacrylates promoted monomer inter-penetration into the tissues and were then polymerized *in situ*. The copolymers in the subsurface could intermediate dental tissues and adhesives. The mixture of the copolymer and the tissue is called a hybrid of tooth substrate and artificial material.

In adhesion to enamel, it had been considered that resin polymerized on the etched enamel could be mechanically bound to the enamel by tag formation. However, addition of a biocompatible monomer into the resin improved the acid resistance of the enamel surface by formation of the hybrid. This hybrid could cover the enamel surface completely and block the permeation of acids into the tooth and protect it against caries. However, the bond strength was not increased by the hybridization.

Bond strength to dentin was improved by the addition of a biocompatible methacrylate into common methacrylates. Surface pretreatments with either acids or EDTA also had a large influence on the bond strength. Ferric ions are required during pretreatment to remove the smeared layer. Phosphoric acid etching gave adverse effects on the bonding to dentin even in the presence of Fe^{3+}.

The hybrid on the tooth structure was confirmed by SEM and TEM observations of the adhered interface after several treatments. Raman spectra also suggested that there was the hybrid on the subsurface of dentin.

Several experiments supported the observation that Fe^{3+} improved the permeability of demineralized dentin subsurface. The structure of dentinal peptides seemed to change the permeability of demineralized dentin. The diffusion rate of the monomers in the tissues is also controlled by the biocompatibility and concentration of the functional monomers. The diffused monomers must be polymerized for good bonding.

1. INTRODUCTION

Hybridization of natural tissues with artificial materials gives good bonding between them. Problems at the juncture between artificial materials and natural tissues such as infection, detachment, loosening, secondary caries, etc., could be minimized. It is also difficult to prepare good adhesives which can bind natural tissues. They would not grow into artificial materials. There are few

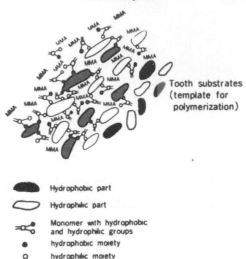

Figure 1. Speculative illustration of inter-penetration and adsorption of monomers with hydrophobic and hydrophilic groups into/onto tooth substrates [1].

ceramics which have a good affinity with bone. A shunt for dialysis patients has changed to an A–V fistula due to the lack of durability.

Artificial materials have been used in the dental field to fix defects because tooth substrates do not regenerate. They have had to be attached to teeth by friction or mechanical retention. Adhesion was impossible. The effectiveness of methacrylates (having both hydrophilic and hydrophobic groups) in adhesion to wet tooth and ivory was suggested after several studies of the interaction of reactive methacrylates with hydroxyapatite. There are questions whether chemical reaction can take place on the solid state within 10 min in the presence of water at 37°C and whether reaction products can stick to the substrates to afford strong connection. A speculative illustration (Fig. 1) shows the production of a hybrid structure between adhesives and the tissue by inter-penetration and adsorption of lipophilic monomers into and/or onto the substrates [1] and several observations to support the mechanism proposed have been obtained.

2. METHACRYLATES TO PROMOTE INTER-PENETRATION AND THEIR POLYMERIZATION

Several methacrylates which promote inter-penetration of monomers into tooth substrates have been prepared (Fig. 2) [2–6]. They possess an aromatic group and a hydrophilic group such as −OH, −COOH or >P(O)OH. Good bonding between tooth substrates and acrylic rods takes place by adding them to methyl methacrylate (MMA). The monomer mixture was polymerized with either a partially oxidized tri-*n*-butyl borane (TBB), benzoylperoxide (BPO)-*N*,*N*-dimethyl *p*-toluidine (DMPT)-sodium *p*-toluenesulfinate (PTSNa) or BPO–PTSNa in the presence of PMMA fine powder. TBB was the effective catalyst to give high bond strength to dentin (see later). Graft polymerization of MMA onto collagen

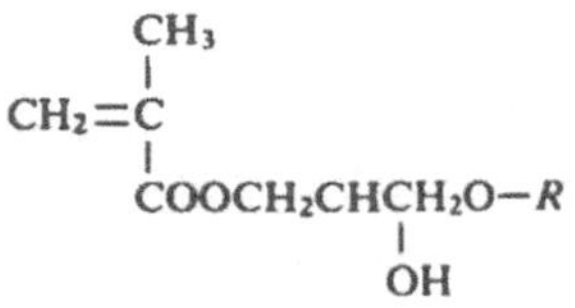

R = Phenyl   2-Hydroxy-3-phenoxypropyl methacrylate (HPPM) [2]
   β-Naphthyl 2-Hydroxy-3-β-naphthoxypropyl methacrylate (HNPM) [3]

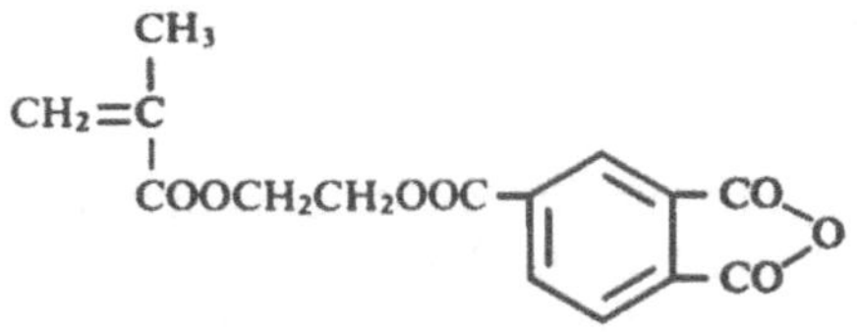

4-Methacryloxyethyl trimellitate anhydride (4-META) [4]

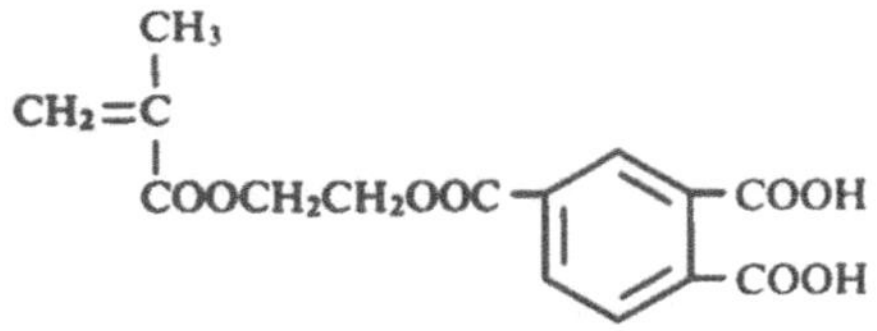

4-Methacryloxyethyl trimellitate (4-MET) [4]

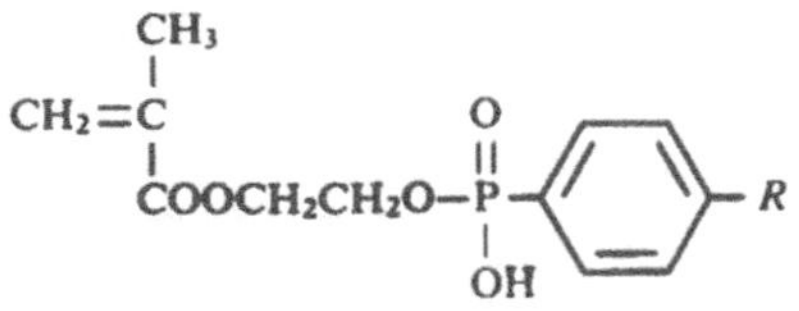

R = H   2-Methacryloxyethyl phenyl phosphoric acid (Phenyl-P) [5]
   CH_3O 2-Methacryloxyethyl p-methoxyphenyl phosphoric acid (CH_3O Phenyl-P) [6]
   Cl   2-Methacryloxyethyl p-chlorophenyl phosphoric acid (Cl Phenyl-P) [6]
   CH_3 2-Methacryloxyethyl p-methylphenyl phosphoric acid (CH_3 Phenyl-P) [6]
   NO_2 2-Methacryloxyethyl p-nitrophenyl phosphoric acid (NO_2 Phenyl-P) [6]

Figure 2. Methacrylates with hydrophobic and hydrophilic groups [2–6].

was believed essential to get good bonding to ivory [7]. However, recent studies suggest that the grafting is not the single mechanism for the bonding [8]. Monomer composition, the polymerization catalyst and dentinal peptides are important in order to get a high bond strength. Diffusion of monomers, which is controlled by the monomer and the substrate, forms the hybrid of the natural tissue and the copolymer after their polymerization.

3. THE HYBRID ON ENAMEL

The hybrid structure on enamel has not been identified directly. The enamel is mainly fine crystals of hydroxyapatite. The interpenetration of monomers is

Table 1.
Relationship between additives in MMA-TBB resin and
length of HCl insoluble part (sum of hybrid and tag)
prepared on phosphoric acid etched enamel [10]

Additive	Length (μm)
none	10
3% HNPM	12
5% Phenyl-P	16
5% 4-META	23

slower in enamel than that in dentin. The thickness of the hybrid was less than
1 μm and was difficult to observe by SEM. Mogi [9] suggested that HCl insoluble structures on enamel etched and applied adhesives became longer at etched
interprismatic substance in the cases of mixtures of MMA and biocompatible
monomers such as HNPM, Phenyl-P and 4-META (Table 1). Inter-penetration
of monomers was promoted by the additives into etched enamel [10]. Maeda
et al. [11] found that acid permeability of enamel etched and applied 4-MET/
MMA–TBB resin which was removed completely was lower than conventional
resins [11]. The enamel treated with the resin did not accept artificial caries
attack. Polymerized resin on etched enamel was removed by Soxhlet extraction
with acetone which could dissolve poly(MMA–co-4-MET) and the enamel was
then soaked in acid, but it was not changed. This phenomena suggested that the
enamel surface was completely covered with an acid impermeable membrane,
a hybrid of the enamel and poly(MMA–co-4-MET). The enamel surface after
the Soxhlet extraction was found to be the same as the acid etched enamel
surface by SEM observation. The hybrid is effective in the prevention of caries.

4. HYBRID ON DENTIN

Graft polymerization of methacrylate onto collagen was considered to be
important [7], and hydrophilic and hydrophobic monomers were not initially

Table 2.
Relationship between etching condition, biocompatible monomers in MMA-
TBB resins and bond strength to etched dentin

Additive	Etchant[a]	Bond strength (MPa)	Reference
none	10-3	13.3 ± 4.9	13
5% HPPM	10-3	10.4 ± 3.8	14
3% HNPM	10-3	10.5 ± 4.0	15
	EDTA 3-2	20.9 ± 2.8	16
5% 4-META	10-3	17.3 ± 4.9	15
	1-1	16.3 ± 0.2	13
	EDTA 3-2	15.7 ± 2.4	17

[a] 10-3 for 30 s; EDTA 3-2 for 60 s.

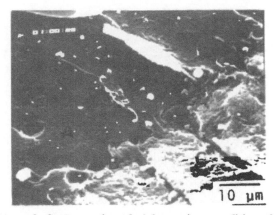

Figure 3. SEM picture of a fracture surface of a joint specimen parallel to tubules showing mushroom shaped cleavage presumably due to polymerization shrinkage (× 2000). (C) tags [12].

effective at bonding to dentin. Cleaning of the ground dentin was necessary for bonding. However, higher-order structures of collagen also had an influence on the bond strength between the 4-META/MMA–TBB resin and cleaned dentin [12]. During studies of the bonding mechanism, inter-penetrtion of monomers to form a hybrid of polymerized methacrylate and dentin on the subsurface of the dentin was found important for the bonding. The functional monomers promote the diffusion of monomers into dentin and increase bond strength to dentin (Table 2) [13–17].

Biocompatible monomers inter-penetrated into dentin and polymerized could be confirmed by the following evidence. Figure 3 shows a SEM picture of a fractured surface parallel to the tubules. There are mushroom shaped cracks on the top of tubules. The cracks were formed by biocompatible monomers which had a strong affinity for tubular wall components. This tooth substrate compatibility with the monomer was stronger than the cohesive energy of adhesive paste. So the curing paste in tubules pulls in the more superficial paste to compensate for the volume contraction in the tubules due to polymerization shrinkage [12]. When the sample shown in Fig. 3 was soaked in 6N HCl for 30 s, the dentin part was demineralized and Fig. 4 shows the results obtained. There is a thin band (B) just below the interface which has resisted acid demineralization. Inter-penetration of monomers into the substrates has taken place soon after the application of 4-META/MMA–TBB resin to the etched dentin and the polymerization is initiated *in situ* with TBB. Figure 4 shows that there are tags (C) in the hybrid (B) and the surface smoothness is similar to that of the cured resin (A) [8]. Figure 4 changed dramatically to Fig. 5 by soaking the sample into 0.01 N sodium hypochlorite for 10 days. The hybrid (B in Fig. 4) was decomposed by sodium hypochlorite and tags (C) were exposed out. This phenomenon suggests that the main constituent of the hybrid is collagen which can be degraded by oxidation [18]. Figure 6 shows a TEM picture of the adhered interface between dentin etched with 10-3 and 4-META/MMA–TBB resin. The

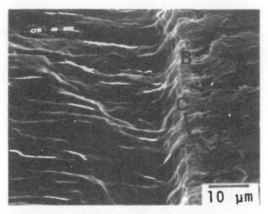

Figure 4. A partially demineralized fracture surface perpendicular to the adhesive interface (× 2000) [8]. The interface is at the right of the porous band (B) which is an acid resistant zone, hybrid of dentin and the adhesive. Tags (C) are seen in the hybrid. (A) is the cured 4-META/MMA-TBB resin.

joint sample could be sectioned without further embedding because the hybridization was a kind of new embedding process which was finished within only 10 min at room temperature. (A) in Fig. 6 is a cured resin. (B) is the hybrid and (C) is a tag [19].

The bond strength to dentin which had been etched with an aqueous solution of 10% citric acid and 3% ferric chloride (10-3) with 4-META/MMA–TBB resin was as high as 18 MPa, and fracture surfaces after tensile tests showed cohesive failure of the cured adhesive. However, the bond strength decreased to 6 MPa when dentin had been etched with either 10% citric or 65% phosphoric acid and adhesive failures were observed (Table 3) [13]. The comparison shows that pure acid etching gives adverse effects on the bonding. Dentinal collagen is a stable peptide and is believed not to change its structure during acid etching

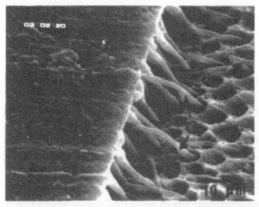

Figure 5. SEM picture of a partially degraded surface of Fig. 4 with 0.01N NaOCl for 10 days (× 2000) [18]. Tags (C) embedded in the hybrid, (B) in Fig. 4, were exposed by the dissolution of the hybrid.

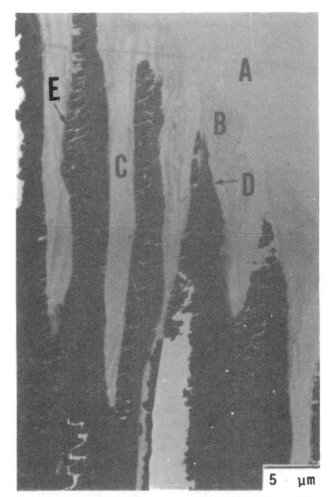

Figure 6. TEM picture of the adhesive interface between 4-META/MMA-TBB resin and dentin etched with 10-3 for 30 s [19]. (A) Cured 4-META/MMA-TBB resin, (B) hybrid of the cured resin and dentin etched, (C) tag, (D) intertubular dentin, (E) peritubular dentin.

with phosphoric acid. The relationship between the pretreatment of collagen and the rate of digestion with trypsin was measured, and it was found that collagen treated with pure acids was hydrolyzed faster than the original. On the other hand, the peptide treated with 10-3 solution did not change the rate of digestion (Fig. 7) [20]. The decrease of bond strength by the phosphoric etching could be reduced by glutaraldehyde pretreatment of dentin [21]. Heat treatment higher than 60°C also decreased the bond strength (Table 4) [20]. The structural change during the heat treatment could be protected by ferric chloride and the bond strength did not change so much. These observations suggest that the higher-order structure of collagen is very important in the adhesion to dentin with 4-META/MMA-TBB resin. There are several reports that ferric chloride

Table 3.
Tensile adhesive strength between human tooth and PMMA rod with 4-META/MMA-TBB resin (MPa) [13].

Pretreatment (%, 30 s)			Substrates (B: bovine)	
Phosphoric acid	Citric acid	$FeCl_3$	enamel	dentin
0	0	0	4.5(B)	2.4 ± 1.8
				7.4 ± 3.4(B)
65	0	0	13.0 ± 1.6	—
0	30	0	10.0 ± 1.3	—
	(60 s)			
20	0	0	—	6.3 ± 1.5
10	0	0	—	5.3 ± 1.4
0	10	0	10.7 ± 4.9(B)	6.3 ± 2.3
				5.3 ± 1.4(B)
0	3	0	—	2.9 ± 0.5
0	10	1	14.0 ± 3.2(B)	15.8 ± 6.1
0	10	3	14.0 ± 2.4(B)	17.5 ± 5.3
				18.1 ± 3.2(B)
0	1	1	8.8 ± 1.9(B)	13.4 ± 2.2
				16.3 ± 0.2(B)

After immersion in water for a day at 37°C.

strengthens the mechanical properties of the cured adhesive and higher bond strength to dentin was obtained [22–24]. An aqueous solution of 1% citric acid and 1% ferric chloride (1-1) was also effective for dentin bonding but not for the enamel.

The 10-3 solution was good for enamel but a little too strong for dentin and milder etchant was desired. Ethylenediamine tetraacetic acid (EDTA) is a good chelating agent and is effective in bonding. EDTA solution at pH 7.4 does

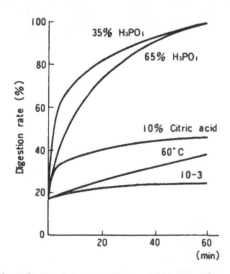

Figure 7. Rate of digestion of treated dentin collagen with trypsin [20].

Table 4.
Effect of heating on the bond strength to bovine dentin heated and then etched with 10-3 [20]

Condition (°C)	Period (min)	Tensile bond strength (MPa)
60 in water	0	18.1 ± 1.4 (5.3 ± 1.4)[a]
	15	7.5 ± 1.5
	30	6.6 ± 0.8
60 in aq. FeCl₃	15	18.1 ± 5.0 (18.3 ± 5.0)[a]
	30	14.4 ± 4.9 (15.8 ± 1.5)[a]

[a] Etched with 10% citric acid for 30 s.

not produce adverse effects on the structure of collagen. However, 0.5 M EDTA etching did not work so well as 10-3 and the bond strength to the dentin was 8 MPa. The bond strength was improved by mixing EDTA Fe salt. The data is shown in Table 5. A half molar EDTA at pH 7.4, a mixture of 0.3 M EDTA·3Na and 0.2 M EDTA·Fe·Na (EDTA 3-2), was the best etchant [17, 25]. As the structure of collagen is not changed with EDTA at pH 7.4, the effectiveness of ferric ions in the EDTA on the bonding cannot be considered the same as that in the 10-3 solution. It is understandable that the ion promotes the polymerization of 4-META/MMA–TBB resin and a higher bond strength is observed. However glutaraldehyde treatment of dentin etched with 0.5 M of pure EDTA sodium salt (EDTA 5-0) gave strange results. The aldehyde diluted with ethanol was effective but was not effective when diluted with water. The mechanical properties of the adhesive is not improved by the aldehyde and that of etched dentin is changed by crosslinking. The diffusion of the aldehyde into dentin etched with EDTA 5-0 is worthy of remark so as to understand the strange phenomena. The diffusion is slower than that into dentin etched with EDTA 3-2. The diffusion rate of the monomers is also an important matter for the formation of the hybrid.

Table 5.
Effect of pretreatment with EDTA on the bond strength to dentin [17][a]

EDTA·3NA / EDTA·Fe·Na (mol)	Tensile bond strength[b] (MPa)
5/0	7.8 ± 1.0
4/1	9.6 ± 4.1
3/2	15.7 ± 2.4
2/3	15.0 ± 2.1
1/4	13.7 ± 5.3
0/5	11.1 ± 1.7

Mean ± SD, n = 3–6.
[a] Joints were soaked in water at 37°C for one day before testing.
[b] Tensile bond strength of 5% 4-META/MMA-TBB resin to bovine dentin demineralized with EDTA for 60 s.

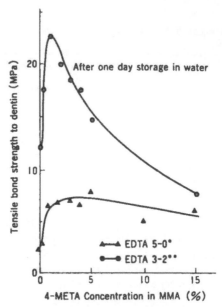

Figure 8. Effect of 4-META concentration in MMA-TBB resins on the bond strength to bovine dentin etched with EDTA for 60 s [17].

The diffusion rate is altered with the concentration of 4-META in MMA. The relationship between the concentration of 4-META in MMA and the bond strength to bovine dentin etched with EDTA 3-2 and 5-0 for 60 s was studied (Fig. 8). The effect of 4-META concentration was clear in EDTA 3-2 and little in 5-0. The ferric ions in the EDTA solution improve the permeability of dentin. The stability of the hybrid against HCl was compared. When the hybrid contains higher amounts of the copolymer it is more stable. Ferric ions and 4-META improved the stability [17].

The depth of demineralized dentin is also important for good bonding. When dentin is demineralized deeper, the monomers cannot reach to the bottom of demineralized dentin before polymerization. Table 6 shows the relation of the etching period and the bond strength to dentin etched with a 10-3 solution [26]. The demineralized dentin not filled with the polymer could easily accept dissolution of collagen in water [27]. The destruction between the hybrid layer and the native dentin due to tensile load also occurred here as shown in Fig. 9, of which dentin had been etched for 60 sec. Normally etching periods are shorter than 30 s and cohesive fracture in the cured adhesive has been observed [12]. When the demineralized depth is thin, a high rate of inter-penetration is not required. Table 7 suggests that 1% 4-META is enough for the dentin etched with EDTA 3-2 [17] and 3% 4-META is required for that treated with 10-3 [14]. 4-META was converted to 4-MET spontaneously when applied to wet dentin [28]. The effect of the etched depth on the bond strength was also observed in HNPM/MMA–TBB resin as shown in Table 2.

Raman microprobe spectroscopy of the hybrid prepared in the subsurface of

Table 6.
Relationship between etching condition, amount of leached Ca^{2+} and tensile bond strength to dentin etched with 4-META/MMA-TBB resin [26].

Composition of pretreatment solution		Treatment time (s)	Amount of leached Ca^{2+} ($\mu g/cm^2$)	Tensile adhesive strength (MPa)
Citric acid (%)	$FeCl_3$ (%)			
10	3	1	54	13.5 ± 2.1
		5	68	9.7 ± 1.2
		10	70	9.4 ± 5.0
		15	74	11.4 ± 4.0
		30	123	12.3 ± 6.0
		60	142	5.1 ± 1.7
5	1.5	1	19	13.4 ± 4.9
		30	41	8.7 ± 5.9
1	1	1	22	11.1 ± 3.6
		30	48	14.4 ± 4.9

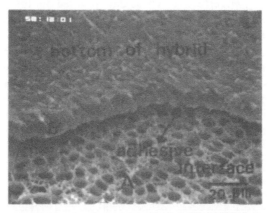

Figure 9. SEM picture of fracture surface on acrylic rod after tensile test of a joint specimen between 4-META/MMA-TBB resin and dentin etched with 10-3 for 60 s showing cohesive failure in dentin (upper) and adhesive failure (lower) (× 1000). (A) cured resin, (B) hybrid [26].

dentin etched with 10-3 by 5% 4-META/MMA–TBB resin was reported by Kato *et al.* [28]. It was a mixture of poly(MMA–co-4-MET) and dentin (Fig. 10). The thickness was 5 μm and the resin content decreased with the depth (Fig. 11). 4-MET was concentrated in the hybrid more than 5% and the data suggested that the monomer had an affinity with dentin. EPMA analysis was also tried and supported the preparation of the hybrid between 4-META/MMA–TBB resin and the dentin, and the thickness was 5 μm in dentin etched with 10-3 [14].

5. ADHESION TO DENTIN WITH PHOSPHORIC METHACRYLATES AND THEIR HYBRIDIZATION

Methacryloxyethyl phosphoric acids with a hydrophobic group were prepared and evaluated to find their effectiveness on the bonding to tooth substrates. The

Table 7.

Effect of concentration of 4-META (4-MET) in MMA–TBB resins and etching condition of dentin on the bond strength to dentin

Concentration	Tensile bond strength (MPa) Etchant	
	10-3 [14]	EDTA 3-2 [17]
0	12.7 ± 6.8	14.1 ± 2.4
0.5	16.2 ± 5.5	17.5 ± 6.4
1	21.8 ± 3.3	22.7 ± 1.6
3	24.5 ± 5.4	18.6 ± 4.1
5	16.8 ± 2.7	15.7 ± 2.4
10	14.7 ± 5.6	7.7 ± 3.8

4-MET for 10-3 (30 s), and 4-META for EDTA 3-2 (60 s).

prepared monomers are illustrated in Fig. 2. They are effective for the adhesion. The phosphoric monomers could promote adhesion without TBB and phosphoric acid etching was effective. These data pointed out that graft polymerization of monomers onto collagen and the importance of ferric ion on the bonding to dentin had to be revised.

The bond strength to dentin with self-curing acrylic resin polymerized by BPO–DMPT (not TBB) in the absence of ferric ions is summarized in Table 8. The dentin was etched with 0.3 M EDTA (pH 7.4) and then applied with a bonding agent of 5% phosphoric methacrylate in MMA initiated by BPO and PTSNa before adhesion [6]. Figure 12 is a hybrid prepared by this adhesive [29]. The hybrid can be formed not only by 4-META/MMA–TBB. Several acids were

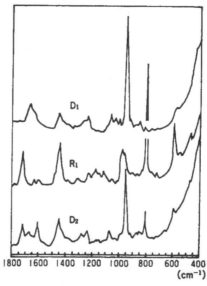

Figure 10. Raman spectra of dentin (D_1), cured 4-META/MMA-TBB resin (R_1) and hybrid prepared on dentin etched with 10-3 for 30 s (D_2) [28].

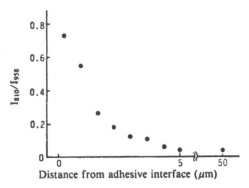

Figure 11. Relationship between ratio of absorption band at 810 nm (characteristic in R_1) and that at 958 nm (characteristic in D_1), and distance from adhesive interface [28].

also effective for the etching of dentin as summarized in Table 9 [30]. These etchants were not so good for the bonding with 4-META/MMA–TBB resin as mentioned already. Some phosphoric monomers were dissolved in MMA and dentin etched with phosphoric acid was joined with a PMMA rod directly by them initiated by BPO-DMPT-PTSNa [31]. The relationships between phosphoric methacrylates in MMA and their bond strength to dentin etched with phosphoric acid or 0.3 M EDTA are shown in Fig. 13 [30, 31]. Bonding of phosphoric monomers/MMA–TBB resins to dentin etched with 10-3 was also studied (Table 10) [32].

6. CONCLUSION

Biocompatible methacrylates with both hydrophobic and hydrophilic groups, which promote the inter-penetration of monomers into the hard tissues, are required for the hybridization of the tissues and polymers. The hybrid is effective to bind them and gives resistance against acids to the tissue surface.

Table 8.
Tensile bond strength of bonding agents containing *p*-substituted Phenyl-P to bovine tooth with BPO-DMPT self-curing resin [6].

	Tensile bond strength (MPa)	
p-Substituent	enamel	dentin
H	12.4 ± 5.1	7.1 ± 1.4
CH$_3$O	12.2 ± 2.6	12.7 ± 2.2
Cl	12.7 ± 1.2	6.3 ± 1.0
CH$_3$	11.0 ± 2.8	10.0 ± 2.2
NO$_2$	12.6 ± 2.7	5.1 ± 1.6
4-META	14.5 ± 2.5	6.1 ± 1.5
none	6.7 ± 1.1	0.3 ± 0.4

Bonding agent was 5% additive in MMA initiated by BPO-PTSNa. Enamel was etched with 65% H$_3$PO$_4$ for 30 s and dentin was etched with 0.3M EDTA (pH 7.4) for 60 s.

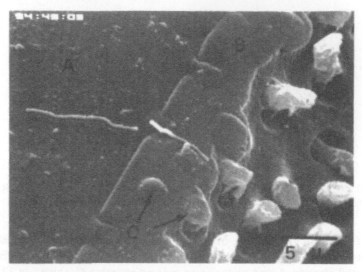

Figure 12. A partially demineralized fracture surface perpendicular to the adhesive interface between dentin., etched with 10-3 for 30 s and applied a bonding agent containing 5% CH$_3$O Phenyl-P in MMA initiated by BPO-PTSNa, and BPO-DMPT self-curing resin ($\times$4000) [29]. The surface was etched with 6N HCl for 20 s. (A) cured resin, (B) hybrid, (C) resin tag.

Table 9.
Effect of pretreatment of dentin on the bond strength to dentin [30]

Pretreatment	Etching period (s)	Tensile bond strength (MPa)
none	–	0.8 ± 0.3
0.3M EDTA (pH 7.4)	60	12.7 ± 2.2
EDTA 3-2	60	10.1 ± 1.2
10% citric acid	30	11.6 ± 3.6
10-3	30	10.0 ± 3.5
65% H$_3$PO$_4$	30	12.2 ± 3.9
0.1N HCl	30	9.3 ± 4.1
Alumina (0.05 μm) polished		3.4 ± 1.9

Bonding agent: 5% CH$_3$O Phenyl-P in MMA (0.5% BPO) and 2% PTSNa in ethanol.

Table 10.
Relationship between *p*-substituents of Phenyl-P, the concentration and tensile bond strength of their MMA–TBB resins to dentin etched with 10-3 [32].

	Tensile bond strength (MPA) Concentration (%)		
p-Substituent	0.5	2.0	5.0
H	17.4 ± 4.5	13.9 ± 5.3	13.1 ± 5.9
Cl	8.3 ± 2.4	13.4 ± 3.0	10.5 ± 4.9
CH$_3$	9.7 ± 2.6	5.2 ± 1.8	7.8 ± 3.3
CH$_3$O	8.9 ± 2.3	12.7 ± 2.2	7.2 ± 7.0
NO$_2$	4.1 ± 1.2	9.2 ± 2.5	7.6 ± 2.8

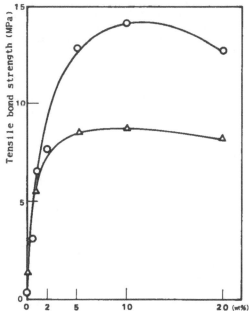

Figure 13. Effect of concentration of phosphoric methacrylates in MMA on the bond strength to dentin etched. (△); Dentin was etched with 65% H_3PO_4 for 30 s and joined with a PMMA rod by BPO-DMPT-PTSNa self-curing resin containing Phenyl-P in MMA [31]. (○); Dentin was etched with 0.3 M EDTA (pH 7.4) for 60 s, then a bonding agent was applied, CH_3O Phenyl-P in MMA containing BPO and PTSNa, and joined with a PMMA rod by BPO-DMPT self-curing resin [30].

REFERENCES

1. Nakabayashi, N., Biocompatibility and promotion of adhesion to tooth substrates. *CRC Crit. Rev. Biocompatibility* **1**, 25–52 (1985).
2. Abe, Y. and Nakabayashi, N., Comparison of bond strength to dentin and enamel with the adhesion promoting monomers, HPPM, HNPM and 4-META. Relationship between interpenetration and bond strength. *J. Jpn. Soc. Dent. Mater. Devices* **4**, 106–111 (1985).
3. Nakabayashi, N., Masuhara, E., Mochida, E. and Ohmori, I., Development of adhesive pit and fissure sealants using a MMA resin initiated by a tri-*n*-butyl borane derivative. *J. Biomed. Mater. Res.* **12**, 149–165 (1987).
4. Takeyama, M., Kashibuchi, N., Nakabayashi, N. and Masuhara, E., Studies of self-curing resins (17). Adhesion of PMMA with bovine enamel or dental alloys. *J. Jpn. Soc. Dent. Appar. Mater.* **19**, 179–185 (1978).
5. Yamauchi, J., Nakabayashi, N. and Masuhara, E., Adhesive agents for hard tissue containing phosphoric acid monomers. *ACS Polym. Preprints* **20**, 594–595 (1979).
6. Nakabayashi, N. and Kanda, K., Synthesis of phosphoryl compounds and their adhesives to bovine tooth. *Kobunshi Ronbunshu* **44**, 91–96 (1988).
7. Masuhara, E., Über die Chemie eines neuen haftfähigen Kunststoff-Füllungs Materials. *Dtsch. Zahnärtl. Z.* **24**, 620–628 (1969).
8. Nakabayashi, N., Resin reinforced dentin due to infiltration of monomers into the dentin at the adhesive interface. *J. Jpn. Soc. Dent. Mater. Devices* **1**, 78–81 (1982).
9. Mogi, M., Study on the application of 4-META/MMA–TBB resin to orthodontics. Adhesion to human enamel. *J. Jpn. Orthodont. Soc.* **41**, 260–271 (1982).
10. Nakabayashi, N., Takeyama, M., Kojima, K., Miura, F. and Masuhara, E., Studies on self-curing resins (22). Adhesion of 4-META/MMA–TBB resin to enamel. *J. Jpn. Soc. Dent. Appar.* **23**, 88–92 (1982).

11. Maeda, M., Mogi, M., Miura, F. and Nakabayashi, N., The adhesion mechanism of MMA resin containing 4-MET to dental enamel. *J. Jpn. Soc. Dent. Mater. Devices* 7, 478–487 (1988).
12. Nakabayashi, N., Takeyama, M., Kojima, K. and Masuhara, E., Studies on self-curing resins (20). Adhesion mechanism of 4-META/MMA-TBB resin to dentin. *J. Jpn. Soc. Dent. Appar. Mater.* 23, 34–39 (1982).
13. Nakabayashi, N., Takeyama, M., Kojima, K. and Masuhara, E., Studies on self-curing regins (19). Adhesion of 4-META/MMA-TBB resin to pretreated dentin. *J. Jpn. Soc. Dent. Appar. Mater.* 23, 29–33 (1982).
14. Abe, Y., Relation between interpenetration and bond strength to dentin with 4-META, Phenyl-P, HNPM, and 4-MET/MMA-TBB resins. *J. Jpn. Soc. Dent. Mater. Devices* 5, 839–851 (1986).
15. Nakabayashi, N., Yamashita, S., Kojima, K. and Masuhara, E., Studies on dental self-curing resins (21). Function of new monomers promoting adhesion to tooth substrates. *Rep. Inst. Med. Dent. Eng.* 15, 37–43 (1981).
16. Shimizu, H., unpublished (1988).
17. Shimizu, H., Adhesion of 4-META/MMA-TBB resins to EDTA treated dentin. The effect of EDTA Fe Na on pretreatment. *J. Jpn. Soc. Dent. Mater. Devices.* 6, 23–36 (1987).
18. Mizunuma, T. and Nakabayashi, N., Adhesion mechanism of 4-META/MMA-TBB resin to glutaraldehyde treated dentin. Analysis of resin reinforced dentin. *J. Jpn. Soc. Dent. Mater. Devices* 3, 642–647 (1984).
19. Nakabayashi, N. and Watanabe, A., SEM and TEM observation of dentin surface treated for adhesion. *Rep. Inst. Med. Dent. Eng.* 17, 45–55 (1983).
20. Mizunuma, T., Relationship between bond strength of resin to dentin and structural change of dentin collagen during etching. Influence of ferric chloride to structure of the collagen. *J. Jpn. Soc. Dent. Mater. Devices* 5, 54–64 (1986).
21. Mizunuma, T. and Nakabayashi, N., Adhesion of 4-META/MMA-TBB resin to dentin modified with formaldehyde and glutaraldehyde. *Jpn. J. Conserv. Dent.* 27, 675–679 (1984).
22. Bowen, R. L., Adhesive bonding of various materials to hard tooth tissues. XXII. Effect of cleanser, mordant, and poly-SAC on adhesion between a composite resin to dentin. *J. Dent. Res.* 60, 809–814 (1980).
23. Jedrychowsky, J. R., Caputo, A. A. and Prola, J., Influence of a ferric chloride mordant solution on resin–dentin retention. *J. Dent. Res.* 60, 134–138 (1981).
24. Kadoma, Y., Kojima, K. and Masuhara, E., Studies on dental self-curing resins. XXV. Effect of ferric chloride or cupric chloride on the polymerization of methyl methacrylate. *J. Jpn. Soc. Dent. Mater. Devices* 2, 495–501 (1983).
25. Shimizu, H. and Nakabayashi, N., Adhesion of 4-META/MMA-TBB resin to bovine tooth substrates treated with EDTA. *Jpn. J. Conserv. Dent.* 28, 270–276 (1985).
26. Kiyomura, M., Abe, Y. and Nakabayashi, N., Effect of pretreatment on the bond strength to dentin. The composition and period. *Jpn. J. Conserv. Dent.* 28, 277–284 (1985).
27. Sasazaki, H., Adaptation of adhesive composite resin to dentinal wall. 1. Bond strength and appearance of gap. *Jpn. J. Conserv. Dent.* 28, 452–478 (1985).
28. Kato, H., Wakumoto, S. and Suzuki, M., Chemical analysis of interface between resin and dentin by Raman microprobe. *J. Jpn. Soc. Dent. Mater. Devices* 5, 232–237 (1986).
29. Nakabayashi, N. and Kanda, K., Hybrid membrane on tooth substrate. *J. Chem. Soc. Jpn., Chem. Ind. Chem.* 1987, 2222–2225 (1987).
30. Nakabayashi, N. and Kanda, K., Bonding agent containing $CH_3O\text{-}OH$ monomer and the dentin pretreating agents. *J. Jpn. Soc. Dent. Mater. Devices* 6, 396–402 (1987).
31. Yamauchi, J., Study of dental adhesive resin containing phosphoric acid methacrylate monomer. *J. Jpn. Soc. Dent. Mater. Devices* 5, 144–154 (1986).
32. Kiyomura, M., Kanda, K. and Nakabayashi, N. *J. Jpn. Soc. Dent. Mater. Devices* 6, 719–726 (1987).

Multiphase Biomedical Materials, pp. 105–114 (1989)
T. Tsuruta and A. Nakajima (Eds)
1989 VSP.

Chapter 7

Mechanical properties of a new type of glass–ceramic for prosthetic applications

T. KOKUBO

Institute for Chemical Research, Kyoto University, Uji, Kyoto-Fu 611, Japan

Summary—A multiphase glass–ceramic, A–W, precipitating apatite and wollastonite in a glassy matrix can form a tight chemical bond with living bone and has a high mechanical strength. Mechanical properties of the glass–ceramic were investigated in terms of its microstructure. Fracture strength of the parent glass was increased only slightly with the precipitation of the apatite $[Ca_{10}(PO_4)_6(O,F_2)]$ alone, but remarkably with the precipitation of the wollastonite $(CaO \cdot SiO_2)$ in addition to the apatite. The increase in the strength was attributed to the increase in the fracture toughness. The wollastonite might effectively inhibit the straight propagation of cracks on the fracture, causing an increase in the fracture surface energy and thereby the fracture toughness. The magnitude of the mechanical fatigue, i.e. the decrease in the fracture strength with loading, of the parent glass in a simulated body fluid was reduced only slightly with the precipitation of the apatite alone, but remarkably with the precipitation of the wollastonite in addition to the apatite. This means that the wollastonite also effectively suppresses slow crack growth due to stress-induced corrosion. When a bending stress of 65 Mpa is continuously applied in the simulated body fluid, glass–ceramic A–W was estimated to be able to withstand over 10 years whereas the parent glass and the glass–ceramic containing apatite alone fail in 1 min. When glass–ceramic A–W was placed in the simulated body fluid without being loaded, it showed an increase in fracture strength. Practical life-times can be expected to be much more prolonged than those estimated above. The glass–ceramic actually implanted into subcutaneous tissue of rat showed little change in fracture strength even after 12 months. As a result, it can be concluded that glass–ceramic A–W is a promising bioactive material for artificial bone usable even under load-bearing conditions.

1. INTRODUCTION

Some inorganic materials such as $Na_2O–CaO–SiO_2–P_2O_5$ glasses (Bioglass®), apatite-containing glass–ceramics (Ceravital®) and hydroxyapatite ceramics are known to form tight chemical bond with living bone and are already used clinically for alveolar ridge reconstruction tooth and middle ear bone implants, etc. [1]. The mechanical strengths of these bioactive materials hitherto known are, however, not sufficiently high. Their uses have, therefore, been limited to applications where strength is a small factor.

The present author and his coworkers recently revealed that a glass–ceramic containing apatite and wollastonite, A–W [2, 3], shows a high bioactivity [2, 4–8] as well as a high mechanical strength. Animal and clinical experiments for application of this glass–ceramic to artificial bones under load bearing conditions are now being conducted [9]. In this study, the mechanical properties of glass–ceramic A–W were investigated in terms of microstructure [10–12].

2. EXPERIMENTAL PROCEDURE

2.1. Sample preparation

Four kinds of glass and glass–ceramics including glass–ceramic A–W were used for the experiments. They have the same composition of MgO 4.6, CaO 44.7, SiO_2 34.0, P_2O_5 16.2, CaF_2 0.5 wt.%, but are composed of different phases as shown in Table 1. G is a glass. It was prepared by pouring a melt of the nominal composition described above on to a stainless steel plate and pressing it into plates 2–6 mm thick. A is a glass–ceramic containing oxy-fluoroapatite $[Ca_{10}(PO_4)_6(O,F_2)]$ as a crystalline phase. It was prepared by heating glass plate G up to 870°C at a rate of 60°C/h and holding it at that temperature for 4 h. A–W is a glass–ceramic containing the oxy-fluoroapatite and β-wollastonite $(CaO \cdot SiO_2)$ as crystalline phases. A–W–CP is a glass–ceramic containing whitlockite $(\beta$-$3CaO \cdot P_2O_5)$ as well as apatite and the wollastonite as crystalline phases. A–W and A–W–CP were prepared by heating a powder compacts of glass G up to 1050 and 1200°C, respectively, at a rate of 60°C/h and holding it at the respective temperatures for 4 h. The glass powder compacts were prepared by crushing glass plate G into powders of particle size below 5 μm and pressing the powders with 1 wt.% paraffin into rectangular bars $25 \times 25 \times 100$ mm under a hydrostatic pressure of 200 MPa.

The samples thus prepared were annealed at appropriate temperatures to remove thermal stresses.

The crystal contents in the samples were determined by a powder X-ray diffraction method using reagent grade CaF_2 as an internal standard. Hydroxyapatite sintered at 1200°C (Mitsubishi Mining and Cement Co.), naturally occurring β-wollastonite (Nichika Co.) and whitlockite prepared by heating a powder mixture of $CaCO_3$ and $Ca_2P_2O_7$ at 1000°C were used as references.

All the samples given in Table 1 had been already shown to form tight chemical bonds with living bone [4–6].

2.2. Measurement of mechanical properties

Fracture strength was measured in air, a dry nitrogen gas atmosphere and a simulated body fluid by a three point bending method using rectangular

Table 1.
Constituent phases of the samples used for the experiments

| Sample | Phase (wt.%) | | | |
	oxy-fluoroapatite	β-wollastonite	whitlockite	glassy phase
G	0	0	0	100
A	38	0	0	62
A–W	38	34	0	28
A–W–CP	20	43	14	23

Table 2.
Ion concentrations of simulated body fluid and human blood plasma

	NA^+	K^+	Mg^{2+}	Ca^{2+}	Cl^-	HCO_3^-	HPO_4^{2-}	
Simulated fluid	142.0	5.0	1.5	2.5	148.8	4.2	1.0	mM
Human plasma	142.0	5.0	1.5	2.5	103.0	13.5	1.0	

specimens of 5 × 5 × 20 mm. The surface of the specimens were finished with no. 2000 alumina powder and their edges were very lightly bevelled with emery paper to minimize the effect of edge flaws. The span length was 16 mm. The loading rate was varied in the range from 0.005 to 2.0 mm/min. The ion concentrations of the simulated body fluid was adjusted so that they are almost equal to those of the human blood plasma [13], as shown in Table 2. The fluid was buffered at pH 7.25 with 50 mM trishydroxymethyl-aminomethane $[(CH_2OH)_3CNH_2]$ and 45 mM hydrochloric acid (HCl). The temperature of the fluid was maintained at 36.5 ±0.5°C. Some samples were subjected to the measurement after being soaked in the simulated body fluid or implanted into subcutaneous tissue of rat for various periods. In the latter case, specimens of 5 × 5 × 25 mm were used. At least six measurements were made to obtain one data point.

Fracture toughness was measured by a double torsion method [14] in a dry nitrogen gas atmosphere using a plate specimen of 3 × 20 × 70 mm having a groove 0.5 mm wide and 1 mm deep at the center of the bottom. Three measurements were made for each sample.

Young's modulus and Poisson's ratio were measured by a resonance method [15] using a cubic specimen of 4 × 4 × 4 mm.

3. RESULTS AND DISCUSSION

3.1. Mechanical properties in an inert environment

Results of the measurement of fracture strengths in air and dry nitrogen gas atmosphere, fracture toughness, Young's modulus and Poisson's ratio are shown in Table 3. The fracture strengths were obtained at a loading rate of 0.5 mm/min. It can be seen from Table 1 that the fracture strengths of the glass in both air and dry nitrogen gas atmosphere is increased only slightly with the precipitation of the apatite alone, but remarkably with the precipitation of the

Table 3.
Fracture strength, σ_{air} and σ_{N_2}, in air and dry N_2 gas, fracture toughness, K_{IC}, Young's modulus, E, and Poisson's ratio, v, of the samples

Sample	σ_{air} (MPa)	σ_{N_2} (MPa)	K_{IC} (MPa m$^{1/2}$)	E (GPa)	v
G	72 ± 25	120 ± 20	0.8 ± 0.1	89	0.28
A	88 ± 12	141 ± 26	1.2 ± 0.1	104	0.27
A-W	178 ± 20	215 ± 26	2.0 ± 0.1	117	0.27
A-W-CP	213 ± 17	243 ± 18	2.6 ± 0.1	124	0.26

T. Kokubo

Table 4.
Flaw size, c, and fracture surface energy, γ, of the samples

Sample	c (μm)	γ (Jm^{-2})
G	28	3.3
A	46	6.4
A–W	55	15.9
A–W–CP	72	25.5

wollastonite in addition to the apatite. The strengths of glass–ceramics A–W and A–W–CP are comparable to or higher than that of the human cortical bone [16].

The fracture toughness is also increased only slightly with the precipitation of the apatite alone, but remarkably with the precipitation of the wollastonite in addition to the apatite. The toughness of glass–ceramics A–W and A–W–CP are also comparable to that of the human cortical bone.

Generally, fracture strength, σ_{IC}, of a ceramic in an inert environment is related to fracture toughness K_{IC} and critical flaw size C as follows [17]

$$\sigma_{IC} = \frac{K_{IC}}{YC^{1/2}},\tag{1}$$

where Y is a constant and is 1.26 for a semicircular surface flaw [18]. Substituting σ_{N_2} and K_{IC} values in Table 3 for σ_{IC} and K_{IC} in eqn. (1), critical flaw sizes given in Table 4 are obtained for the respective samples. It can be seen from Tables 3 and 4 that the increase in the fracture strength with the precipitation of the wollastonite is attributed to the increase in the fracture toughness, but not to the decrease in the flaw size.

Fracture toughness of a ceramic is generally related to Young's modulus E, Poisson's ratio v and fracture surface energy γ as follows [19]

$$K_{IC} = \left[\frac{2E\gamma}{(1 - v^2)} \right]^{1/2}\tag{2}$$

Substituting K_{IC}, E and v values in Table 3 for K_{IC}, E and v in eqn. (2), fracture surface energies given in Table 4 are obtained for the respective samples. It can be seen from Tables 3 and 4 that the increase in the fracture toughness with the precipitation of the wollastonite is attributed to the increase in the fracture surface energy rather than the increase in Young's modulus and not to the increase in the Poisson's ratio.

The fracture surface energy includes not only the energy to make a newly cleaved surface, γ_s, but also energies needed to create an irregular fracture surface $C\gamma_s$, crack branching γ_b, plastic deformation γ_p and others γ_o, as follows [20]

$$\gamma = C\gamma_s + \gamma_b + \gamma_p + \gamma_o.\tag{3}$$

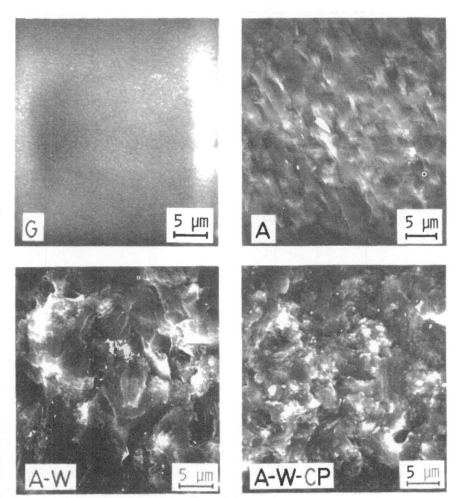

Figure 1. SEM pictures of fracture surfaces of the samples.

Contributions from the plastic deformation and the others are negligible for normal ceramics near room temperature. The increase in the fracture surface energy with the precipitation of the wollastonite is then attributed to the increase in the energies needed to create an irregular fracture surface and crack branching. This is confirmed on scanning electronmicrographs of the fracture surface of the samples, which are shown in Fig. 1. It can be seen from Fig. 1 that the irregularity of the fracture surface is increased remarkably with the precipitation of the wollastonite. The wollastonite might effectively inhibit straight propagation of cracks, causing them to turn or branch out.

3.2. *Mechanical properties in a simulated body environment*

Fracture strengths in the simulated body fluid are shown in Fig. 2 as a function of stressing rate. The fracture strengths in dry nitrogen gas atmosphere are

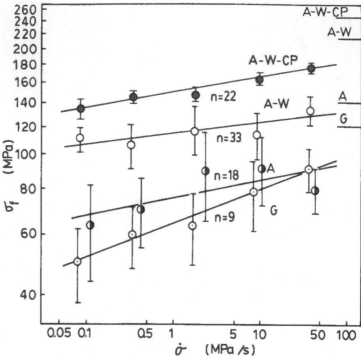

Figure 2. Fracture strength σ_f in the simulated body fluid as a function of stressing rate $\dot\sigma$. Bars at the right hand side show the strength in dry N_2 gas atmosphere.

shown for comparison at the right hand side of Fig. 2. It can be seen from Fig. 2 that all the samples examined show a fracture strength in the simulated body fluid which is lower than that in the inert environment and decreases with decreasing stressing rate. This means that all the examined samples show mechanical fatigue, i.e. decrease in the strength with loading, in the simulated body fluid, as other ceramics in an aqueous environment. The magnitude of the decrease in the strength with decreasing stressing rate, however, largely depends upon the kind of the samples. The magnitude is the largest for glass G. It is reduced only slightly with the precipitation of the apatite alone, but remarkably with the precipitation of the wollastonite in addition to the apatite. This means that the wollastonite effectively also suppresses slow crack growth due to stress-induced corrosion [17] in the simulated body fluid. The precipitation of the whit-lockite gives an adverse effect on the mechanical fatigue of the glass–ceramic.

Generally, the dependence of the fracture strength σ_f of a ceramic in aqueous environment upon stressing rate $\dot\sigma$ is represented by the following equation [21]

$$\ln \sigma_f = \left(\frac{1}{n+1}\right) \ln \dot\sigma + \left(\frac{1}{n+1}\right) \ln [B(n+1)] + \left(\frac{n-2}{n+1}\right) \ln \sigma_{IC}, \qquad (4)$$

where n is a constant related to the velocity of the slow crack growth and B is a constant related to n and the fracture toughness. Applying eqn. (4) to the

Table 5.
n and ln B values of the samples

Sample	n	ln B
G	9	−5.74
A	18	−0.063
A–W	33	8.53
A–W–CP	22	2.09

relations in Fig. 2, and substituting σ_{N_2} in Table 3 for σ_{IC}, n and ln B values given in Table 5 are obtained.

The life-time t_f, i.e., time to failure, of the ceramic showing the mechanical fatigue represented by eqn. (4) is given by the following equation [21]

$$\ln t_f = \ln B + (n - 2) \ln \sigma_{IC} - n \ln \sigma_a, \tag{5}$$

where σ_a is an applied stress. Substituting n and ln B values in Table 5 and σ_{N_2} values in Table 3 for n, ln B and σ_{IC} in eqn. (5), we obtain the life-times shown in Fig. 3 as a function of applied stress for the respective samples. It can be seen from Fig. 3 that glass–ceramics A–W and A–W–CP show much longer life-times than glass G and glass–ceramic A. For example, when a bending stress of 65 MPa is continuously applied in the simulated body fluid, glass–ceramics A–W and A–W–CP can withstand over 10 years whereas glass G and glass–ceramic A fail in 1 min. The bending stress of 65 MPa for an artificial bone corresponds to that of 200 MPa or larger for the natural bone, because a load-bearing cross-sectional area of the former is 3 times that of the latter or larger, if the former takes a solid rod form as usual, instead of a hollow cylindrical form of the latter.

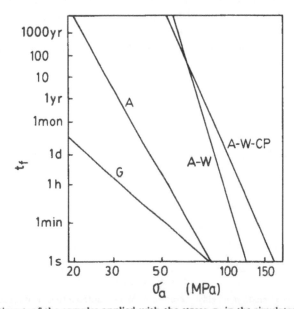

Figure 3. Life-time t_f of the samples applied with the stress σ_a in the simulated body fluid.

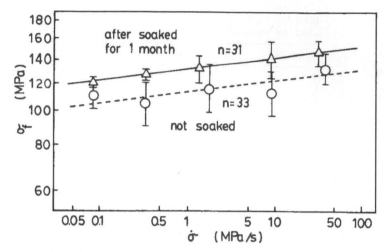

Figure 4. Fracture strengths σ_f of glass–ceramic A–W soaked in the simulated body fluid for 1 month as a function of stressing rate $\dot{\sigma}$, in comparison with those of the glass–ceramic before soaking.

The life-times estimated above are those for the materials continuously loaded. Actual artificial bone is not always loaded in a body. Figure 4 shows fracture strengths in the simulated body fluid of glass–ceramic A–W soaked in the fluid for 1 month, in comparison with those of the glass–ceramic before soaking. It can be seen from Fig. 4 that glass–ceramic A–W shows an increase in strength when placed in the simulated body fluid without being loaded. The dependence of the fracture strength upon the stressing rate is affected little by soaking. This result indicates that a much longer life-time than that estimated above can be expected in the practical case. The increase in strength with soaking might be attributed to crack blunting which is caused by apatite deposition on the glass–ceramic in the simulated fluid [22].

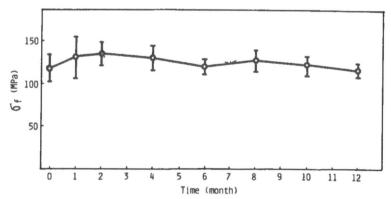

Figure 5. Fracture strength σ_f of glass–ceramic A–W implanted into subcutaneous tissue of rat for various months.

3.3. Mechanical properties in actual body environment

Figure 5 shows the fracture strength of glass–ceramic A–W which had been implanted into subcutaneous tissue of rat for various months. The measurements were made in the simulated body fluid at a loading rate of 0.5 mm/min. It can be seen from Fig. 5 that glass–ceramic A–W shows little change in the fracture strength even after being implanted in the body for 12 months. This might be a result of cancellation of the increase in the strength due to the crack blunting with the decrease in the strength due to slow crack growth.

4. CONCLUSION

It can be concluded from the above experimental results and discussion that glass–ceramic A–W is a promising bioactive material for artificial bone, usable even under load-bearing conditions.

Acknowledgements

It is a pleasure to thank my colleagues. Experimental works on mechanical properties were conducted by Dr S. Ito and Mr M. Shigematsu, Institute for Chemical Research, Kyoto University. Mr T. Kitsugi of Faculty of Medicine, Kyoto University performed *in vivo* experiments. Glass–Ceramics A–W and A–W–CP were prepared by Mr T. Shibuya and Mr M. Takagi, Nippon Electric Glass Company. The encouragement and valuable suggestions of Professor T. Yamamuro and Professor M. Tashiro as well as useful comments of Professor S. Sakka are gratefully acknowledged.

REFERENCES

1. Hulbert, S. F., Bokros, J. C., Hench, L. L., Wilson, J. and Heimke, G., Ceramics in clinical applications, past, present and future. In: *Ceramics in Clinical Applications*, Ed. Vincenzini, P., Elsevier, Amsterdam, 1987, pp. 3–27.
2. Kokubo, T., Shigematsu, M., Nagashima, Y., Tashiro, M., Nakamura, T., Yamamuro, T. and Higashi, S., Apatite- and wollastonite-containing glass–ceramic for prosthetic application. *Bull. Inst. Chem. Res., Kyoto Univ.* 60, 260–268 (1982).
3. Kokubo, T., Ito, S., Sakka, S. and Yamamuro, T., Formation of a high-strength bioactive glass–ceramic in the system $MgO–CaO–SiO_2–P_2O_5$. *J. Mater. Sci.* 21, 536–540 (1986).
4. Nakamura, T., Yamamuro, T., Higashi, S., Kokubo, T. and Ito, S., A new glass–ceramic for bone replacement; evaluation of its bonding to bone tissue. *J. Biomed. Mater. Res.* 19, 685–698 (1985).
5. Kitsugi, T., Yamamuro, T., Nakamura, T., Higashi, S., Kakutani, Y., Hyakuna, K., Ito, S., Kokubo, T., Takagi, M. and Shibuya, T., Bone bonding behavior of three kinds of apatite-containing glass–ceramic. *J. Biomed. Mater. Res.* 20, 1295–1307 (1986).
6. Kitsugi, T., Nakamura, T., Yamamuro, T., Kokubo, T., Shibuya, T. and Takagi, M., SEM-EPMA observation of three types of apatite-containing glass–ceramics implanted in bone; the variance of a Ca-P-rich layer. *J. Biomed. Mater. Res.* 21, 1255–1291 (1987).
7. Kokubo, T., Hayashi, T., Sakka, S., Kitsugi, T. and Yamamuro, T., Bonding between bioactive glasses, glass–ceramics or ceramics in a simulated body fluid. *Yogyo-Kyokai-Shi* 95, 785–791 (1987).
8. Kitsugi, T., Yamamuro, T., Nakamura, T., Kokubo, T., Takagi, M., Shibuya, T., Takeuchi, H. and Ono, M., Bonding behavior between two bioactive ceramics *in vivo. J. Biomed. Mater. Res.* 21, 1109–1123 (1987).
9. Yamamuro, T., Nakamura, T., Higashi, S., Kasai, R., Kakutani, Y., Kitsugi, T. and Kokubo, T., Artificial bone for use as a bone prosthesis. In: *Progress in Artificial Organs—1983, Vol. 2*, Eds Atsumi, K. *et al.*, ISAO Press, Cleveland, 1984, pp. 810–814.

10. Kokubo, T., Ito, S., Shigematsu, M., Sakka, S. and Yamamuro, T., Mechanical properties of a new type of apatite-containing glass–ceramic for prosthetic application. *J. Mater. Sci.* **20**, 2001–2004 (1985).

11. Kokubo, T., Ito, S., Shigematsu, M., Sakka, S. and Yamamuro, T., Fatigue and life time of bioactive glass–ceramic A–W containing apatite and wollastonite. *J. Mater. Sci.* **22**, 4067–4070 (1987).

12. Kitsugi, T., Yamamuro, T., Nakamura, T., Kakutani, Y., Hayashi, T., Ito, S., Kokubo, T., Takagi, M. and Shibuya, T., Aging test and dynamic fatigue test of apatite–wollastonite-containing glass–ceramics and dense hydroxyapatite. *J. Biomed. Mater. Res.* **21**, 467–484 (1987).

13. Gamble, J., *Chemical anatomy, physiology and pathology of extracellular fluid*, 6th Edn, Harvard University Press, Cambridge, 1967.

14. Williams, D. and Evans, A., A simple method for studying slow crack growth. *J. Test. Eval.* **1**, 264–70 (1973).

15. Hirao, K. and Soga, N., Cryostat for semiautomatic measurement of heat capacity and elastic moduli between 1.6 and 400 K. *Rev. Sci. Instrum.* **54**, 1538–42 (1983).

16. Curry, J., The mechanical properties of bone. *Clin. Orthopaed. Related Res.* **73**, 210–231 (1970).

17. Davidge, R., *Mechanical Behavior of Ceramics*. Cambridge University Press, London, 1979.

18. Irwin, G., Crack extension force for a part-tough crack in a plate. *J. Appl. Mech.* **29**, 651–654 (1962).

19. Lawn, B. and Wilshaw, T., *Fracture of Brittle Solids*. Cambridge University Press, London, 1975.

20. Soga, N., Elastic moduli and fracture toughness of glass. *J. Non-crystalline Solids* **73**, 305–313 (1985).

21. Frieman, S., Fracture mechanics of glass. In: *Glass: Science and Technology*, Vol. 5, Eds Uhlman, D. and Kreidl, N., Academic Press, New York, 1980, pp. 21–78.

22. Kokubo, T., Hayashi, T., Sakka, S., Kitsugi, T., Yamamuro, T., Takagi, M. and Shibuya, T., Surface structure of a load-bearable bioactive glass–ceramic A-W. In: *Ceramics in Clinical Applications*, Ed. Vincenzini, P., Elsevier, Amsterdam, 1987, pp. 175–184.

Multiphase Biomedical Materials, pp. 115–138 (1989)
T. Tsuruta and A. Nakajima (Eds)
1989 VSP.

Chapter 8

Metallic biomaterials for replacement of hard tissue

ISHI MIURA,[1] HITOSHI HAMANAKA,[1] OSAMU OKUNO[1]
and KENZO ASAOKA[2]

[1]*Institute for Medical and Dental Engineering, Tokyo Medical and Dental University,
2-3-10 Surugadai, Kanda, Chiyodaku, Tokyo 101, Japan*
[2]*Department of Dental Engineering, School of Dentistry, Tokushima University,
Tokushima 770, Japan*

Summary—Metallic biomaterials for replacement of hard tissue, such as artificial bones, hip joints and materials for artificial tooth roots, were examined in the following five areas: 1, porous titanium for dental implants; 2, porous titanium–zirconium alloy; 3, ceramic coated metals; 4, unidirectionally solidified eutectic alloy for the femur head; 5, precise casting of Ni–Ti shape memory alloy for medical and dental uses. Porous pure titanium and Zr–40wt.%Ti alloys were made by the sintering of spherical particles (420–500 μm in diameter) made by the rotating electrode process (REP). They showed suitable pore structure for bone ingrowth and superior biomechanical compatibility. Porous ceramic coating on metals, *in situ* composite by unidirectional solidification of Co–Cr–C eutectic alloys, precise casting of NiTi and titanium alloys were also studied and they seemed to be useful for the development of metallic biomaterials.

1. INTRODUCTION

Metals and ceramics have been used as replacements for human hard tissues. However, both of them have advantages and disadvantages.

Metals have high tensile strength, large elongation and high impact strength. On the other hand, ceramics are strong in compression but their elongation under tensile stress is very poor and their tensile strength is much inferior to that of metals, which make their resistivity to impact force very low.

Biocompatibility is one of the characteristics of ceramics and, therefore, alumina, zirconia, bioglass, appatite, etc., have been investigated or used as a substitute for human hard tissues.

Some precious metals have been used as substitutes for hard tissues for a long time. Recently, non-precious metals such as Co–Cr alloy, 316L stainless steel and Ni–Cr alloy have also been used, but they are considered to be inferior to ceramics in terms of biocompatibility. However, some alloys possess good corrosion resistance and are biocompatible, for example, titanium or zirconium, both in column IV of the periodic table.

Titanium and zirconium have superior mechanical properties compared to ceramics, and are better substitutes for human hard tissues than alumina or other ceramics.

Among the ceramics, hydroxyapatite has a composition similar to that of bones or teeth and has good biocompatibility; moreover, it bonds directly with new bones. However, hydroxyapatite lacks sufficient bending strength, tensile strength and impact resistivity. To improve these properties, hydroxyapatite is used to coat metals such as titanium or its alloys which have good mechnical properties. This field is now being studied by many researchers.

It is well known that porous surface materials allow new bone to grow into their pores, resulting in an anchor effect, and giving firm bonding between biomaterials and bones. It is thought that biomaterials can increase the functions by making their surfaces porous. We have studied such biomaterials for artificial bones, hip joints and tooth roots. Our investigations have focused on the following five areas:
(1) porous titanium for dental implants;
(2) porous titanium–zirconium alloy;
(3) ceramic-coated metals;
(4) unidirectionally solidified eutectic alloy for femur heads; and
(5) precise casting for Ni–Ti shape memory alloy and titanium alloy for dental and medical use.

2. POROUS TITANIUM FOR DENTAL IMPLANT

2.1. Materials and methods
In this section, the results of porous titanium which we have studied are shown [1]. The use of porous materials may overcome some of the problems faced by the clinician. Retention and stabilization have been enhanced by fibrous tissue and bone ingrowth through such implants. Adjustment of the modulus of elasticity of implants close to host bone supports a stable host site environment. The use of porous metals prevent bitterness and subsequent fracture to which ceramics are prone. The surface of titanium coated with titanium oxide is passive and is protected from corrosion and tarnishing. Titanium also has good mechanical properties and excellent biocompatibility. In anticipation of these values, porous titanium dental implants were fabricated and tested.

Titanium has a high melting point and is a very reactive metal. It is difficult to obtain the powder by ordinary atomization. We have tried to make spherical powder grains by the rotating electrode process (REP) [2]. In this procedure, a pure titanium rod was made from sponge titanium using an arc-melting furnace. The titanium rod connected to the anode terminal was in turn connected to a high-speed motor. A tungsten rod was used for the negative pole. The terminals were placed in a vessel evacuated to 1×10^{-3} Pa and with high-purity argon gas later introduced. The anode rod, rotating at high speed, was melted in an argon atmosphere by arc fusion. The melting metal flew about due to centrifugal force and formed spherical powder granules. Particle diameter was controlled by adjusting the rotational speed of the anode rod.

The titanium powder was sifted out. Only particles of 420–500 μm diameter were used. The compacted particles had weak bonding with each other, after sintering with an alumina mold made by the lost wax method, in a 1×10^{-3} Pa vacuum at 1000°C. The presintered specimen was sintered further in a vacuum without the mold for 8.6×10^4 s (24 h) at 1400°C. The specimen after sintering had a height of 10 mm and a diameter of 4 mm.

Porosity and average pore diameter were measured on the cross-section of the specimen using the feature analysis system.

The specimen was polished, and the micro-Vicker's hardness number was measured at the center of the particle before and after sintering. Compressive strength, tensile strength, bending strength and low-cycle fatigue compressive strength were tested using an Instron-type universal testing machine. Fractography was observed by scanning electron microscopy.

The pure titanium rod was inserted into an alumina mold, and the titanium powder was used to fill the space surrounding the core. Sintering was performed under the same conditions as with the fully porous metal specimens. Figure 1 shows the titanium powder, the titanium core and the sintered surface porous implant. The surface porous implant had a height of 10 mm and a diameter of 4 mm. The core had a wide variety of lengths and diameters—0.8, 1.2 and 1.6 mm, respectively. In addition, a surface porous specimen (Fig. 2) was made to determine the bonding strength along the interface of the two materials. The

5 mm

Figure 1. Titanium powder granules of 420–500 μm diameter, titanium core metal, and porous titanium-coated dental implant.

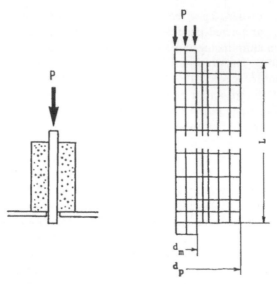

Figure 2. Push-out test method, and finite element analysis model. $d_m = 2.0$ mm.

push-out test was then performed. Bonding strength was determined from the relation between coated length and push-out force.

2.2. Results

The hardness number of the particles fabricated by REP was 140 Hv. After sintering, the hardness of the same part of the particle was 245 Hv.

Since the hardness number was 245 Hv after sintering, the level of solution

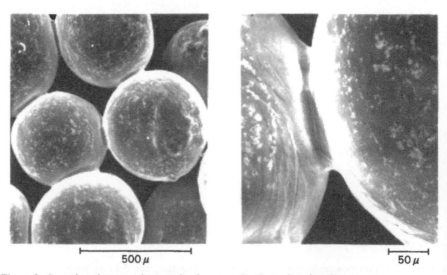

Figure 3. Scanning electron micrograph of porous titanium after sintering.

atoms in the titanium particle was below 0.4% for oxygen and below 0.2% for nitrogen.

Various pore diameters were, of course, observed at the surface of the sintered specimen, as shown in Fig. 3. Loose bonding was also observed on the cross-section. The average diameter of the channel voids was 220 μm. The density of the specimen was 2.6 g/cm and the porosity was 41.5%. The reason why the particles were not able to form a closely packed structure may be due to the differences in the diameters, from 420 to 500 μm, of the packed spherical particles. The specimen then has a porosity lying between that of a body-centered cubic (b.c.c.) type structure and a simple cubic-type structure. The diameter of the connecting void channels for a b.c.c. structure theoretically lies between 0.31 (77 μm) and 0.58 (145 μm). Here, r is the radius of the lattice point particle and is estimated at 250 μm. If the shape of the ingrowth bone can be easily changed, bone of 160 μm diameter can pass the narrowest channel on the closest plane in the b.c.c. structure. However, the diameter of the channel is actually a little less because of the bonding areas of the particles. A single cubic structure is looser than a b.c.c. structure, and bone of 200 μm diameter can easily pass the narrow void.

Klawitter and Hulbert [3] demonstrated experimentally that interconnecting pores of 150 μm diameter provide an optimum setting for the ingrowth of osseous tissue. Predecki *et al.* [4] showed that ingrowth of bone was most rapid in titanium with a pore diameter of 500 μm. Therefore, it might be expected from these results that this porous titanium would promote bone ingrowth and retention of the implant.

The compressive strength after sintering of the fully porous titanium was 184 ± 34 MPa for ten specimens at a cross-head speed of 0.5 mm/min. The shrinkage to ultimate strength was 12 ± 2% and the modulus of elasticity was 5.2–5.5 GPa. An example of the stress–strain curves obtained is shown in Fig. 4. The compressive tests were also carried out at various cross-head speeds from

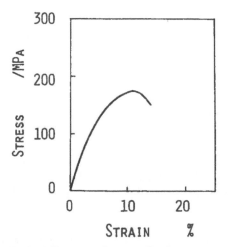

Figure 4. Compressive stress–strain curve of porous titanium

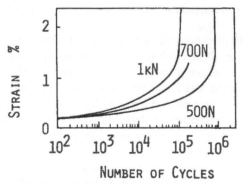

Figure 5. Results of fatigue tests at various loads.

0.5 to 500 mm/min, and there was little difference between the results. Bite frequency is about ten times per second at most. The maximum bite speed is then approximately 1000 mm/min. The mechanical properties of the material do not change at these strain rates.

The bending strength of a specimen of 20 mm length, 4 mm width and 3 mm thickness was 73.0 ± 9.0 MPa. Tensile strength was 45.6 ± 6.4 MPa. Compressive fatigue tests at a cross-head speed of 1.0 mm/min were performed. Figure 5 shows the relation between the number of loaded cycles and shrinkage of the specimen. In the early testing cycles, shrinkage of about 100 μm was observed, but there was little dimensional change after the completion of these cycles. The shrinkage of the specimen after 8×10^5 cycles on 500 N was approximately 1%. After 9×10^5 cycles, the specimen rapidly shrank and fractured. But, on 1 kN, fatigue failure occurred after about 1×10^5 cycles.

Figure 6 shows the falling off surface of the titanium particle after compressive fracture, and the tear surface after fatigue fracture. A river pattern was observed from the fractography of the compressive test (Fig. 6a) and a striation-like marking was observed from the fractography of the fatigue test (Fig. 6c). Figures 6(b) and (d) show the features of the fracturing bonded areas. In the case of the compression test, the bonded area was torn from both surfaces of the particles. The river pattern was marked during the fracture process. The striation after fatigue testing was marked from the next process. In the early cycles of the test, a few cracks were generated in the bonded areas. These cracks may then grow and spread, as shown in Fig. 6(d).

Push-out testing was performed. A compressive load was applied to the core metal as shown in Fig. 2, and bonding strength at the interface was examined. Figure 7 shows the relation between push-out fractured load and porous titanium-coated length. The calculation of the stress distribution using the finite element method is shown in Fig. 8. Here, the moduli of elasticity of the core and of the porous metal were estimated at 118 and 5 GPa, respectively. Point *A*, as shown in Fig. 2, denotes greatly increased stress concentration. Cracks initiated near point *A* may then pass through the interface, resulting in the failure of the specimen. If the shear strength at the core–porous-metal

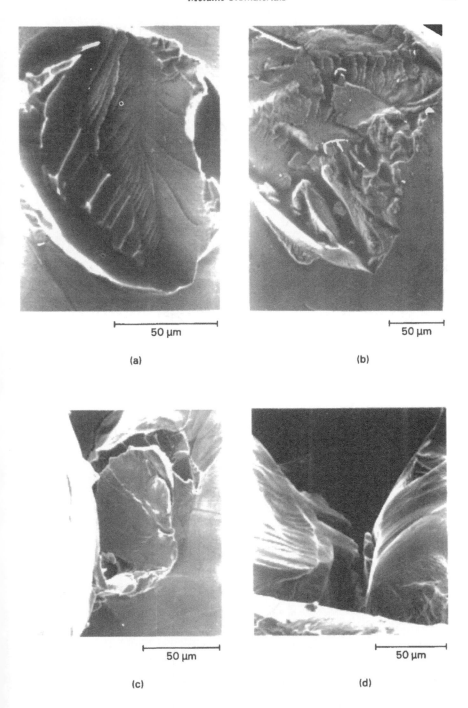

50 μm

(a)

50 μm

(b)

50 μm

(c)

50 μm

(d)

Figure 6. Scanning electron micrograph of porous-titanium-fractured surface. (a) River pattern on compressive fracture surface; (b) features of compressive fracture at the bonding areas; (c) striation-like marking on fatigue fracture surface; (d) features of fatigue fracture at the bonding areas.

I. Miura et al.

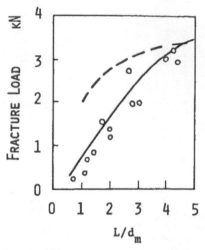

Figure 7. Results of push-out test. (O), experimental data; (---) theoretical results.

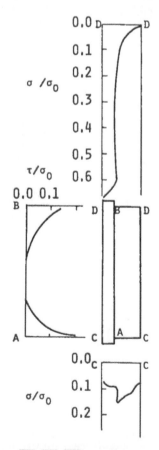

Figure 8. Stress distribution along $\overline{AB}$, $\overline{CC}$, $\overline{DD}$ by push-out test.

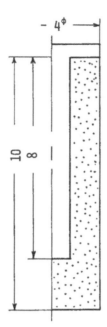

Figure 9. Porous titanium-coated dental implant model.

interface is estimated at 180 MPa from the results of compressive strength testing, push-out force is theoretically the broken line shown in Fig. 7.

The surface porous titanium implant used in the compression test is schematically shown in Fig. 9. The relationship between compressive strength and core length was examined. The core has a diameter of 1.2 mm and a length of 0.0–10.0 mm. The powder immediately under the core greatly increased for every core length used by a factor of 2.1 times stress concentration from the finite element calculations. Table 1 shows the results of compression tests for

Table 1.
Results of the compression tests on the porous titanium-coated dental implants with various core length.

Core length (mm)	Compressive strength (MPa)	Shrinkage (%)
0.0	234	15
0.0	158	9
0.8	175	16
2.0	195	18
2.6	195	15
3.2	189	15
4.9	176	15
5.9	178	15
6.7	205	16
9.4	238	15
10.0	234	15
10.0	159	4

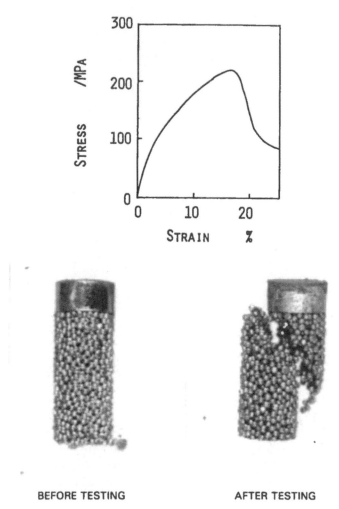

BEFORE TESTING   **AFTER TESTING**

Figure 10. Compressive stress–strain curve of porous titanium-coated dental implant, and before and after testing of a specimen.

specimens of core diameter 1.2 mm. One such specimen, with a core length of 10 mm, was fractured by buckling and had a low compressive strength. Figure 10 shows the stress–strain curve and had a low compressive strength. Figure 10 shows the stress–strain curve and macrofractography of the surface porous titanium dental implant with a core length of 8 mm. The fracture was caused by slippage along 30 μm from the loading direction. The cores of all these specimens were bent by buckling.

The relationship between compressive strength and core thickness was examined just as before. The core has a length of 8 mm and diameters of 0.8, 1.2, and 1.6 mm, respectively. Table 2 shows the results for compressive strength and ductility of specimens with cores embedded in the titanium

Table 2.
Results of the compression tests on the porous titanium-coated
dental implants with various core thickness

Core thickness (mm)	Compressive strength (MPa)	Shrinkage (%)
0.8	237	19
	228	17
1.2	218	16
	228	13
1.6	232	13
	232	12

powder. It is clear from the finite element calculations that there are no remarkable differences between the stress concentrations found under the core among the different thickness used. This is also true for compressive strength.

2.3. Discussion

Various implant materials with a wide variety of modulus of elasticity, from porous PMMA with 0.5 GPa to alumina with 340 GPa, have been used. When occlusive force generates compression on a protruding post, biomechanical compatibility, i.e. suitability of the material to the bone, was studied with finite element simulation. The material with the most desirable modulus of elasticity for dental implants was determined from the results. Figure 11 shows the simulation model. The stress distribution in the model, which was composed of 159 nodes and 144 elements, was calculated using the two-dimensional axisymmetric finite element method. The modulus of elasticity of the alveolar bone was estimated to be 14 GPa, and the same value was applied over the entire alveolar bone region.

Figure 12 shows the relationship between the modulus of elasticity of the implant materials and the displacement of the gingival line under loading. The material with a modulus of under 100 GPa has great shrinkage. In view of the strain mismatch at the interface, it would not seem unreasonable to suppose that implant material of low elastic modulus such as a polymer is of limited use. When occlusive force generates compression on a protruding post, immediately beneath the implanted material in the alveolar bone decreases in compressive stress. The interface has a distribution of shear stress. Points C and D show greatly increased shearing stress concentration; the central part of the interface has almost no shearing stress concentration. Interfacial flaking may then be initiated from the points C and D. Figure 13 shows the relationship between the modulus of elasticity of the material and maximum compressive stresses at the points A and B. The stress values are also divided by the compressive stress on the protruding post. The material with a modulus of over 20 GPa has minimally higher stress concentration at point A than material with a modulus of exactly 20 GPa. However, at point A there is increased stress concentration. Material such as alumina with a high modulus of elasticity is used for implant, so there is greatly increased stress concentration at the part corresponding to point.

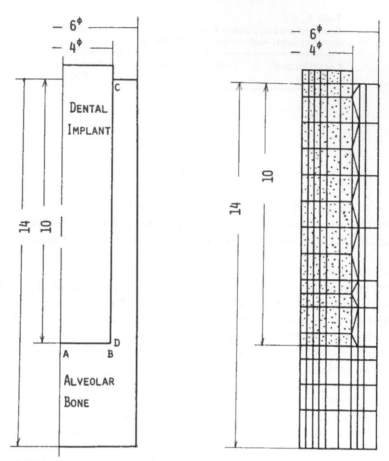

Figure 11. Finite element calculation model and mesh pattern.

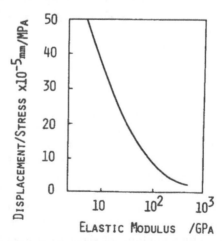

Figure 12. Relationship between the elastic modulus of dental implant material and displacement between points C and D.

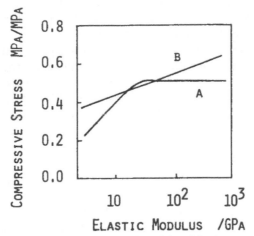

Figure 13. Relationship between the elastic modulus of dental implant material and compressive stress concentration factor at points A and B.

Figure 14 shows the changes in interfacial shear stresses at the points *A* and *B* corresponding to different moduli of elasticity of implanted materials. If the material has a soft elastic modulus, eruption of the interface may be initiated from the neck of the implant. On the other hand, hard material may cause a tear by separating from the bone near the bottom when exposed on loading. Material with a modulus of elasticity of 70–200 GPa is favorable for adhesion to tissue on the grounds of good biomechanical compatibility.

From these results, it is feasible to use surface porous titanium as dental implant material. Material with thick core and with tissue ingrowth into pores

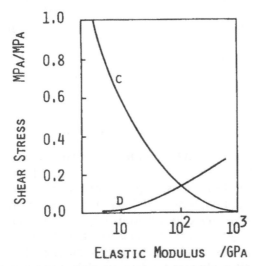

Figure 14. Relationship between the elastic modulus of dental implant material and maximum stress concentration factor at points C and D.

perhaps has an increased modulus of elasticity. Therefore, such material has favorable biomechanical compatibility together with retention and stabilization of the implant, depending on ingrowth of bone.

Another problem commonly observed in dental implant materials is strength; the implants must be strong enough to withstand biting force. A number of studies on biting force have been made. However, it varies markedly between individuals. In the molar region it may range from 370 to 890 N; in the premolar region, from 230 to 450 N; from 140 to 330 N on cuspids and from 88 to 250 N on incisors [5, 6]. The level of compressive stress at the neck of a tooth is about 13–17 MPa, as calculated from the biting force and the area of the neck. The proportional limit of the tooth structure is 125–224 MPa, and its compressive breaking strength is 232–305 MPa [5, 7], this value being 15–18 times that of the stress developed during mastication. Compressive strength of the porous titanium in this study was almost the same value as the proportional limit of the natural tooth. The fatigue strength was 40 MPa, 2- or 3-times the stress developed during mastication. The engineer always multiplies these expected stress values by a 'safety factor' in order to ensure that a particular structure may be able to withstand a certain amount of overstress. In this regard, porous titanium undoubtedly requires higher strength. Pore size, porosity and the strength of the particle-to-particle interface are important factors contributing to the strength of porous materials and improvements are required on these points. On the other hand, porous titanium may be stronger in alveolar bone because of bone ingrowth. The compression test in this study constitutes a stricter inspection than in an oral environment, since the test was performed between ridge jigs.

The possibility of using porous titanium on dental implant can be considered with respect to biomechanical compatibility and strength. Material with an elastic modulus of 70–200 GPa has been used most frequently as dental implant material. Porous titanium has a lower elastic modulus than this value. However, a thick core with ingrowth of tissue into void channels raises the elastic modulus of the material.

It was concluded that a porous titanium dental implant has better biomechanical compatibility than low modulus materials such as polymers and high modulus materials such as ceramics.

3. POROUS TITANIUM–ZIRCONIUM ALLOY

Various porous metals manufactured by the powder metallurgy process have been investigated for the fixation of implants to bone [8]. One of the problems with porous metals is their poor strength as mentioned in the conclusion of the previous section. Open pores at least 150 μm in diameter are needed to admit the greatest amount of bone into porous metals. However, such large open pores lower the strength of porous metals to that of human cortical bone. Due to fatigue stress, porous metals need higher static strength.

The object of this study was to develop high-strength porous zirconium–titanium alloys for implants by sintering a mixture of spherical zirconium and

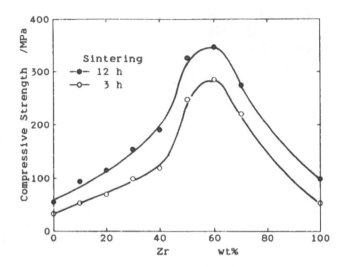

Figure 15. Compressive strength of porous zirconium–titanium alloys.

titanium particles. Both zirconium and titanium possess good mechanical properties, excellent corrosion resistance and are biocompatible. Zirconium and titanium form a complete solid solution and a mixture of these particles can be sintered easily and quickly. The best zirconium/titanium ratio for implant materials was investigated from the standpoints of mechanical properties and pore structure.

The spherical titanium and zirconium particles we used were made by the rotating electrode process (REP). Spherical particles (420–500 μm) were sieved, then mixed and sintered in a high vacuum for 3–12 h at 1400°C.

Compressive strength as a function of zirconium wt.% is shown in Fig. 15. The compressive strength of porous zirconium–titanium alloys showed a peak at 60 wt.% zirconium. The compressive strength of this composition was 4- to 8-times greater than that of porous pure titanium and about 2 times that of human cortical bone. The elastic modulus of this porous zirconium–titanium alloy was 8.3–10.5 GPa. The strain at ultimate compressive stress was more than 7%, and the bending strength of porous Zr–40wt.%Ti alloy was 2- to 3-times greater than that of porous pure titanium.

The cross-section of porous Zr–40wt.%Ti alloy revealed a porosity of 36% and pore size of 210 μm (Fig. 16), sufficient to allow bone to grow into the pores.

The results of this study indicate that porous Zr–40wt.%Ti alloy would be suitable for the fixation of implant to bone, due to its excellent mechanical properties, suitable pore structures for bone in-growth and superior biocompatibility.

4. CERAMIC-COATED METALS [9]

The surfaces of 316L stainless steel and COP-cobalt alloy were coated with ceramic by the plasma-spray method. High-pure alumina, white alumina and

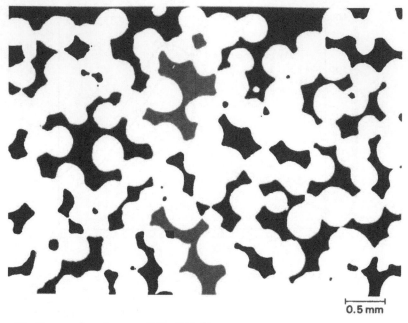

0.5 mm

Figure 16. Cross section of porous 60Zr–40Ti alloy.

gray alumina containing 10 or 40% titania were used as the coating ceramics. To improve the interfacial strength of alumina/metals, undercoating of Ni–Cr alloys was attempted. To obtain a rough surface, mixed alumina and salt powder was plasma-sprayed on the base metals covered with wire net. Interfacial shear strength was measured by the push-out test (Fig. 17). Some specimens were dipped in Ringer solution at 37°C for 6 months. Degradation of these specimens in Ringer solution was determined by measuring the interfacial strength. Fatigue testing was carried out with a hip joint simulator.

The interfacial strength of ceramic-coated metals was greatest with white alumina-coated 316L stainless steel. The interfacial strength of alumina-coated

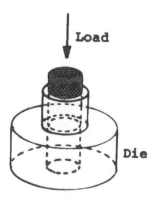

Load

Die

Figure 17. Push-out test.

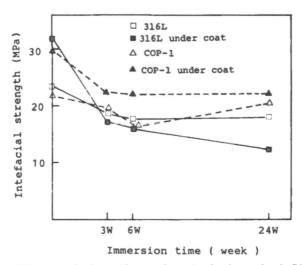

Figure 18. Interfacial strength of ceramic coated metals after immersion in Ringer's solution.

316L stainless steel could be increased by about 30-40% by under-coating with Ni–Cr alloy. However, the long-term immersion test in Ringer solution showed that white alumina-coated 316L stainless steel with Ni–Cr alloy under-coating was degraded considerably (Fig. 18), because of galvanic corrosion that occurred between Ni–Cr and stainless steel. The best result was obtained with white alumina-coated COP-1 alloy, which has high corrosion resistance.

The cyclic load also had considerable influence on the interfacial strength of ceramic-coated metals. After 10 cyclic loadings of 588 kN, the interfacial strength of ceramic-coated metals decreased to 50–70% of the initial interfacial strength and after 10 cyclic loadings, the interfacial strength reached zero. The

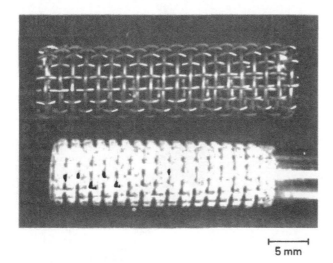

Figure 19. Porous ceramic coated metals.

interfacial strength between ceramics and metals must be improved further for practical use.

Figure 19 shows rough porous coating on metals. Mixed alumina and salt powder was plasma-sprayed on base metals covered with wire net. After plasma spraying, the wire net was removed and the salt was dissolved in water. This porous ceramic coating on metals is suitable for in-growth of bone into the porous structure.

5. UNIDIRECTIONALLY SOLIDIFIED Co–Cr–C EUTECTIC ALLOY FOR FEMUR HEADS

We have been studying the application of metal-based composite materials for medical or dental use for a long time, including the application of *in-situ* composites for the femur head [10]. *In-situ* composites, i.e. unidirectionally solidified eutectic composites, have excellent mechanical properties in the longitudinal direction and good wear resistance, and thus are believed suitable artificial bone substitutes, especially for the femur head.

A eutectic alloy of 56.9% Co–40.9% Cr–2.2% C was solidified unidirectionally using the floating zone melting method in a mold made using the lost-wax process.

Figure 20 shows an artificial femur head upon removal from the mold, and Fig. 21 shows a schematic drawing of the cross section of a undirectionally solidified artificial femur head corresponding to the microstructures in

20 mm

Figure 20. Unidirectionally solidified C73 artificial femur head just taken out from the mold.

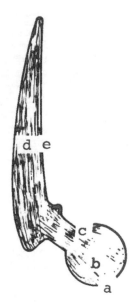

Figure 21. Schematic drawing of the cross section of a unidrectionally solidified artificial femur head. a, b, c, d and e correspond to the microstructures in Fig. 22(a), (b), (c), (d) and (e) respectively.

Fig. 22(a)–(e), respectively. These photographs show that the stem of the head is composed of fibrous structure.

The wear resistance was examined in the femur head by using an artificial joint simulator (Fig. 23) in 0.9% NaCl solution, and the results showed that this alloy was much better than non-directionally solidified Co–Cr–C eutectic alloy.

According to our study, unidirectionally solidified Co–Cr–C eutectic alloy is a very good substitute for femur heads.

6. PRECISE CASTING MACHINE FOR NiTi SHAPE MEMORY ALLOY AND TITANIUM ALLOY FOR DENTAL AND MEDICAL USE [11]

Figure 24 is a schematic diagram of a new casting machine developed during the course of this study. This machine consists of an upper melting chamber and a lower casting chamber, and uses an argon arc vacuum-pressure system. Although the principle is the same as the 'CASTMATIC', the details are different. The main features of the new developments are as follows: (i) The melting and casting chambers are evacuated to a higher degree of vacuum using an oil diffusion pump. The degree of vacuum reaches 1.3×10 Torr (1×10 mm Hg) when there is no mold, and 2.6×10 Torr (2×10 mm Hg) 10 min after setting a mold in the casting chamber. (ii) A heater is set to control the mold temperature in the casting chamber. This is heated at a given temperature prior to setting the mold. Figure 25 illustrates the inside of this casting machine. The heater can be moved up and down by the lever marked with an arrow to save time in heating and cooling the mold. (iii) The copper crucible can be split to

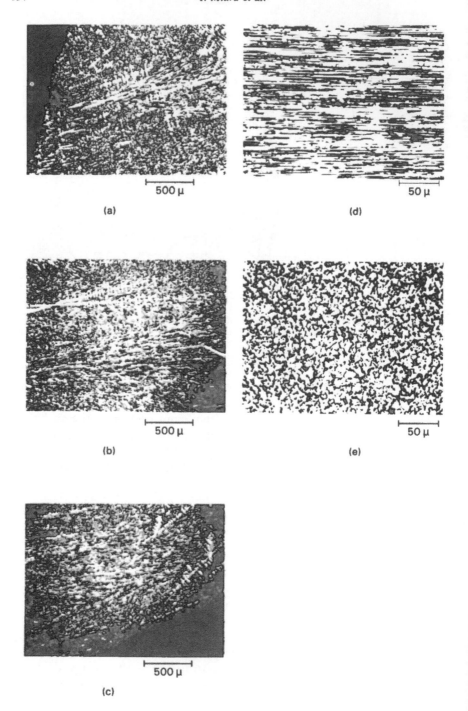

Figure 22. Microstructures of a unidirectionally solidified C73 femur head. (a) Top part of the head ball; (b) central part of the head ball; (c) neck part of the femur head; (d) longitudinal section of the stem; and (e) transverse section of the stem.

Figure 23. Artificial joint simulator.

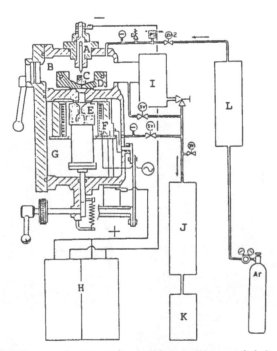

Figure 24. Schematic diagram of a new casting machine. A: Water cooled electrode. B: Melting chamber. C: Metal. D: Copper crucible. E: Mold. F: Heater. G: Casting chamber. H: D.C. electric sources. I: Oil diffusion pump. J: Vacuum tank. K: Rotary vacuum pump. L: Compressed argon gas tank.

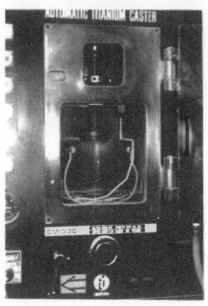

Figure 25. The inside of the casting machine. The heater is movable up and down by the lever marked with an arrow.

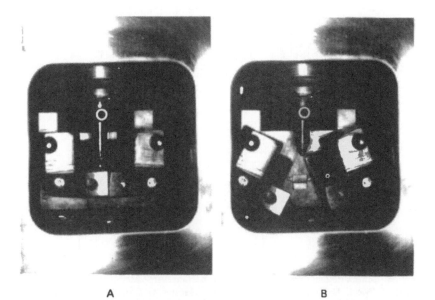

A　　　　　　　　　　　　　　　B

Figure 26. The copper crucible. A: Before casting; B: after casting.

let molten metal drop into the mold. The crucible (A, before casting and B, after casting) is shown in Fig. 26. (iv) A new control system was developed, which is shown in Fig. 27. (v) The vacuum tank and the compressed argon gas tank were set to operate more efficiently. (vi) The volume of material that can be melted was increased by using a water-cooled electrode and double DC electric sources.

Figure 27 shows the casting process using this machine. After a mold and metal are set in the machine, the upper and lower chambers are evacuated. It takes about 10 min to reach the correct degree of vacuum and mold temperature. Then argon gas is fed into the upper chamber by pushing the 'start' button and an electric arc starts automatically at the given pressure. After the alloy

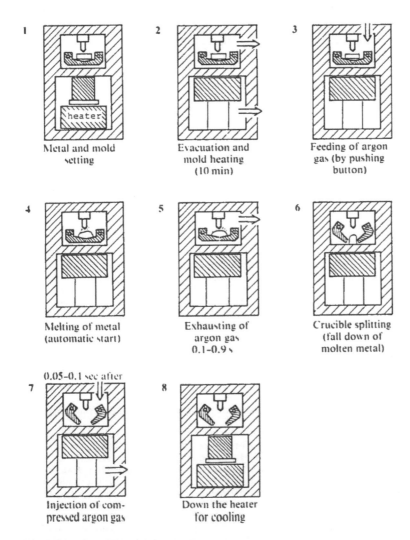

Figure 27. The casting process of this machine.

melts down, the new control system can be started by pushing the 'cast' button. At first the upper chamber is evacuated for 0.1–0.9 sec, then the copper crucible is split to drop the molten metal and, 0.05–0.1 s later, compressed argon gas is injected into the upper chamber. This control system works automatically according to a given program.

Wax patterns were invested in commercial phosphate-bonded silica invest-ment (SHOFU SUMAVEST). In this study, three items were cast: 2-mm diameter and 50-mm long tensile test specimens; 0.3-mm thick 20 × 30 mm sheet; and crowns. The molds were burned out once at 850°C in air and then were heated to 600°C in the machine. Pure titanium (Grade 1), NiTi alloy (49.2 Ti–50.8 Ni and 49.6 Ti–50.4 Ni in at.%) were used for casting. Stress–strain curves were obtained using an Instron testing machine and strain meter. Hard-ness was measured by micro-Vicker's hardness tester and microstructures were observed on the cross section of the castings.

The results and conclusions are as follows. Gas in the mold was removed by heating the mold in a high vacuum, and as a result reactions between the molten metal and molds were much decreased. Titanium and NiTi alloys could be cast even with commercial phosphate bonded silica investment without losing their important properties. Elongation of cast pure titanium was over 30% and NiTi alloy showed super-elasticity, which is an important property for clasps. The new control system developed here proved very useful in preventing internal macro-defects in castings and in improving castability.

Our new casting machine is considered promising for the casting of titanium and NiTi alloys because it allows casting of these alloys with conventional tech-niques and investments.

REFERENCES

1. Asaoka, K., Kuwayama, N., Okuno, O. and Miura, I., Mechanical properties and biocom-patability of porous titanium for dental implants. *J. Biomed. Mater. Res.* 19, 669–773 (1985).
2. Okuno, O. and Miura, I., The properties of Co–Cr–Mo alloy powder prepared by rotating elec-trode process. *Rep. Inst. Med. Dent. Eng., Tokyo Med. Dent. Univ.* 14, 1–5 (1980).
3. Klawitter, J. J. and Hulbert, S. F., Application of porous ceramics for the attachment of load bearing internal orthopedic applications. *J. Biomed. Mater. Res. Symp.* 2, (part 1), 161–229 (1971).
4. Predecki, P., Stephan, J. E., Auslaender, B. A., Mooney, V. L. and Kirkland, K., Kinetics of bone growth into cylindrical channels in aluminium oxide and titanium. *J. Biomed. Res.* 6, 375–400 (1972).
5. Philips, R. W., Physical properties of dental materials, rheology, color, thermal properties and biological considerations. In: *Science of Dental Materials*, Eds Saunders, Philadelphia, 1982, pp. 28–62.
6. Howell, A. H. and Manly, R. S., An electronic strain gauge for measuring oral forces. *J. Dent. Res.* 27, 705–712 (1948).
7. Stanford, J. W., Weigel, K. V., Paffenbarger, G. C. and Sweeney, W. T., Compressive proper-ties of hard tooth tissue. *J. Am. Dent. Assoc.* 60, 746–756 (1960).
8. Okuno, O., Miura, I., Kawahara, H., Nakamura, N. and Imai, H., New porous zirconium-titanium alloys for implants. *Biomaterial '84 Transactions*, pp. 121 (1984).
9. Hamanaka, H., Miura, I. and Yamaguchi, H., Applications for ceramic coating for bio-metals. *Proceedings of Second Conference of Japanese Society for Biomaterials*, pp. 159–162 (1980).
10. Hamanaka, H., Doi, H., Kohno, O. and Miura, I., Dental castings of NiTi alloys. Part 2. New casting techniques for NiTi alloys. *Jpn. J. Dent. Mat.* 4, 573–579 (1985).
11. Hamanaka, H., Doi, H. and Miura, I. The application of unidirectionally solidified Co–Cr–C eutectic composites for artificial hip joint. *Rep. Inst. Med. Dent. Eng., Tokyo Med. Dent. Univ.* 20, 11–18 (1986).

Multiphase Biomedical Materials, pp. 139-152 (1989)
T. Tsuruta and A. Nakajima (Eds)
1989 VSP.

Chapter 9

Use of immobilized enzyme reactors in automated clinical analysis

TAKASHI MURACHI

Department of Clinical Science and Laboratory Medicine, Kyoto University Faculty of Medicine, Kyoto 606, Japan

Summary—One of the important applications of immobilized enzymes to clinical medicine is the application to biochemical analysis in clinical laboratories. There are two principal ways of clinical analytic application of immobilized enzymes: one is in the form of a membrane that constitutes a biosensor with an electrode, and the other is in the form of a column or reactor. A bioreactor for diagnostic use can be constructed using immobilized enzyme-bearing glass beads packed into a minicolumn through which the solution of analyte is made to flow, while the product that emerges from the outlet of the enzyme column is quantified with high sensitivity. We utilized flow injection analysis, coupled with a device for chemiluminometric determination of hydrogen peroxide. We employed several oxidases which produce hydrogen peroxide from each of the analytes in serum and urine, e.g. glucose, uric acid and lactic acid. We have also invented a novel method that permits the determination of urea in terms of hydrogen peroxide using urease, glutamate dehydrogenase and glutamate oxidase all immobilized and aligned in one reactor column in this order from upstream to downstream. The design and practical use of a commercial model of an automated analytic instrument for the four items mentioned above is described.

1. INTRODUCTION

The diagnosis of diseases is heavily dependent on clinical laboratory data. A large number of biochemical analyses on serum and urine samples are to be routinely carried out, and quality control of the results should be high, not only inside an individual laboratory but also among different laboratories. Automation is thus the natural choice of procedure.

Ever since a prototype flow instrument for clinical analysis was introduced in the market [1], a large variety of automated analyzers have been devised which include different types such as those of discrete, continuous-flow and the centrifugal principle. Any modern clinical laboratory now produces perhaps more than 90% of the analytic data by use of these automated instruments.

Almost in parallel with the development of automated instruments, another remarkable change has occurred in clinical laboratories. That is the introduction of enzymes as analytic reagents [2]. According to the survey conducted by the Japanese Medical Association in 1985, covering some 2000 institutions, the share of enzymatic methods reached 83.8% for serum glucose, 98.8% for total cholesterol, 99.9% for triglyceride and 86.1% for urinary glucose. The manufacturing industry for diagnostic enzymes is still growing in Japan as well as in other countries.

Since an enzyme is a biocatalyst, the enzyme reagent can be repeatedly used many times, if such reagent could be separated from other reactants at the end of each cycle of the reaction and recovered from the reaction mixture for the next cycle. The latter situation can be accomplished by employing immobilized enzymes as reagents [3].

In this chapter, the usefulness of enzyme reagents for clinical analysis is discussed, with special reference to the application of immobilized enzyme reactors to automated flow analysis.

2. USE OF ENZYMES IN CLINICAL ANALYSIS

2.1. Enzymes as reagents

Table 1 illustrates the current status of the use of enzyme reagents for clinical analysis. Even though the entries to the table may not be exhaustive, very wide use of a variety of enzymes is apparent. This is quite reasonable because an enzyme reagent is much more specific for the analyte molecule than a conventional chemical reagent, and an enzyme reaction can be carried out under much milder conditions, such as at neutral pH and at room temperature, compared with most of the chemical reactions which require heating and/or making solutions strongly acidic or alkaline. The majority of the items of clinical analysis is for the assays of organic constituents of the serum and urine. The latter fact is also greatly favored by the enzyme reagents, since these constituents are the natural substrates of the respective enzymes.

2.2. Enzymatic determination of inorganic constituents

Even inorganic constituents of the blood and urine can be determined using specific enzyme reagents. For example, magnesium can be determined using hexokinase (HX) or glucokinase (GK) and glucose 6-phosphate dehydrogenase (G6PDH):

$$\text{D-glucose} + \text{Mg.ATP}^{2-} \xrightarrow{\text{HK or GK}} \text{D-glucose 6-phosphate} + \text{Mg.ADP}^{2-}$$

$$\text{D-glucose 6-phosphate} + \text{NADP}^+ \xrightarrow{\text{G6PDH}} \text{6-phospho-D-gluconate}$$

$$+ \text{ NADPH} + \text{H}^+$$

These two coupled reactions are usually used for the determination of glucose, but they can also be utilized for determining magnesium if one makes the supply of magnesium to the medium the limiting factor. The enzymatic method for magnesium determination is very simple and accurate, and can be applied to assays on both serum [4] and urine [5], giving results in excellent agreement with those obtained by atomic absorption spectrophotometry (Fig. 1).

Table 1.
Clinicial analysis using enzymes as reagents

Item	Enzymes as analytic reagents	Final signal[a]
Glucose	Glucose oxidase	H_2O_2
	Pyranose oxidase	H_2O_2
	Hexokinase and glucose-6-phosphate dehydrogenase	NAD(P)H
	Glucose dehydrogenase	NAD(P)H
Cholesterol	Cholesterol oxidase	H_2O_2
	Cholesterol dehydrogenase	NADH
Cholesterol ester	Cholesterol ester hydrolase and cholesterol oxidase	H_2O_2
Triglyceride	Lipase and glycerol dehydrogenase	NADH
	Lipase and glycerol kinase	ADP
	Lipase, glycerol kinase and glycerol-3-phosphate dehydrogenase	NADH
	Lipase and glycerol oxidase	H_2O_2
Phospholipid	Phospholipase C and alkaline phosphatase	P_i
	Phospholipase D and choline oxidase	H_2O_2
Free fatty acid	Acyl-CoA synthetase and myokinase	ADP
	Acyl-CoA synthetase and acyl-CoA oxidase	H_2O_2
Urea	Urease	NH_3
	Urease and glutamate dehydrogenase	NAD(P)H↓
	Urease, glutamate dehydrogenase and glutamate oxidase	H_2O_2
Ammonia	Glutamate dehydrogenase	NAD(P)H↓
	Glutamate dehydrogenase and glutamate oxidase	H_2O_2
Uric acid	Uricase	H_2O_2
Creatinine	Creatininase, creatinase and sarcosine oxidase (or sarcosine dehydrogenase)	H_2O_2 (or NADH)
	Creatinine deiminase	NH_3
Bilirubin	Bilirubin oxidase	O_2↓
Bile acid	3α-Hydroxysteroid dehydrogenase	NADH
Lactic acid	Latate dehydrogenase	NADH
	Lactate oxidase	H_2O_2
Pyruvate	Lactate dehydrogenase	NADH↓
	Pyruvate oxidase	H_2O_2
Sialic acid	Neuraminidase, NANA-aldolase and pyruvate oxidase	H_2O_2
L-Amino acid	L-Amino acid oxidase	H_2O_2
Magnesium	Hexokinase and glucose-6-phosphate dehydrogenase	NADPH
Phosphate (P_i)	Pyruvate oxidase	H_2O_2
Calcium	Phospholipase D and choline oxidase	H_2O_2
NAD(P)H	NAD(P)H oxidase	H_2O_2
ADP	Pyruvate kinase and lactate dehydrogenase	NADH↓
AMP	Myokinase, pyruvate kinase and lactate dehydrogenase	NADH↓
Choline esterase activity	Choline oxidase	H_2O_2
Amylase activity	α-Glycosidase and glucose oxidase	H_2O_2
Asparate transaminase activity	Glutamate dehydrogenase	NADH
	Malate dehydrogenase	NADH↓
	Oxaloacetate decarboxylase and pyruvate oxidase	H_2O_2
Alanine transaminase activity	Lactate dehydrogenase	NADH↓
	Pyruvate oxidase	H_2O_2
	Glutamate oxidase	H_2O_2

[a] Signals mean the production (increase), except for those with downward arrows which mean the decrease.

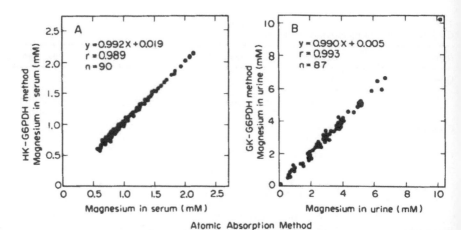

Figure 1. Correlation between the results of determination of magnesium in serum obtained by the HK–G6PDH method (A) or in urine obtained by the GK–G6PDH method (B) and those obtained by the atomic absorption method. Superimposed points are not illustrated. Taken from refs 4 and 5.

As shown in Table 1, some other inorganic constituents including phosphorus [6] and calcium [7] can also be determined using specific enzymes.

3. USE OF IMMOBILIZED ENZYMES

3.1. Immobilization

The principle and methods for immobilization of enzymes have been well documented [8, 9]. There are two principal types of immobilized enzymes used for clinical analysis. One is the 'membrane type', in which the enzyme molecules are entrapped into the matrix of a membrane. The other is the 'reactor type', which is actually a column- or tube-type reactor with enzymes covalently, or similarly firmly, bound to the solid phase inside. We have been mostly using commercially available porous glass beads as the solid particles. These glass beads are derivatized to have alkylamino groups on to which enzyme proteins are covalently bound by the following two methods. (i) The enzyme protein and glass beads are allowed to react with a bifunctional reagent, such as glutaraldehyde, which forms Schiff base linkages with both the primary amino groups (mostly ε-amino groups) of the enzyme protein and with those of the glass beads [10]. (ii) The carbohydrate moiety, if present in an enzyme molecule to be immobilized, is first oxidized with periodate to form dialdehyde groups which are then coupled with the primary amino groups of the glass beads through Schiff base formation [11, 12] (Fig. 2).

3.2. Column-type reactors

As shown in Fig. 3, the membrane type is mostly used in combination with an electrode to constitute a 'biosensor', which has found increasing applications not only to laboratory analysis but also to continuous monitoring, e.g. as a part of an 'artificial pancreas' [13]. The column-type reactor is not widely used at

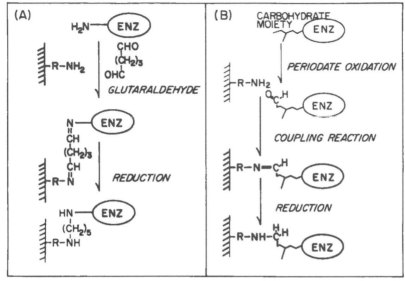

Figure 2. Two principal methods for immobilization of enzymes to the solid phase by covalent linkages. (A) a method using glutaraldehyde; (B) a method using the carbohydrate moiety of the enzyme protein.

present in routine laboratory studies, but its applicability to any continuous-flow analytic system is well known [2, 3]. A minicolumn measuring 0.5–2.0 mm i.d. and 5–40 mm in length is packed with immobilized enzyme-bearing glass beads. Both ends of the packed column are covered with small pieces of Nylon net with a lattice of $40 \times 40 \mu m$ to prevent bed movement that may occur under the pressure pulse caused by pumping solutions in the system during analysis.

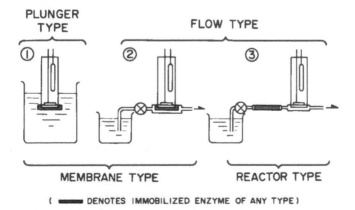

Figure 3. Use of immobilized enzymes for analysis in the form of a membrane (Nos 1 and 2) or column (No. 3). The detector can be an electrode as illustrated, or it can be any one of a number of other sensing systems. Note that solutes diffusion in all direction is maintained in the case of a membrane type, while it is highly restricted in the case of a column type, due to unidirectional flow of the solution through the column.

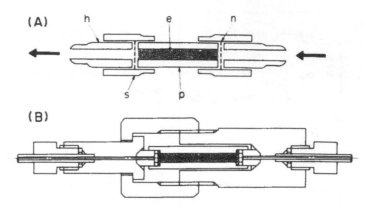

Figure 4. Schematic representation of immobilized enzyme columns. (A) a home-made column (1.0–2.0 mm i.d. × 5–40 mm in length); (B) a commercial design (enzyme column size: 3.0 mm in i.d. × 20 mm in length). e, immobilized enzyme; h, nipple; n, Nylon net; p, plastic tubing; s, silicone rubber tubing. Direction of the flow is from right to left both in (A) and (B).

Figure 4 shows two examples of the column reactors [3]. The column is simply inserted into the flow stream of an analyzer, and the product of the enzymatic reaction that emerges from the outlet of the reactor can be monitored.

4. COMBINED USE OF SEVERAL IMMOBILIZED ENZYMES

When the assay of a given item needs to utilize two or more different enzymes, one can choose the simultaneous or the sequential use of these immobilized enzymes.

4.1. Use of co-immobilized enzymes

Cholesterol in the serum is present both as unesterified (or free) and esterified cholesterol. Therefore, total cholesterol can be determined by utilizing two enzymes, cholesterol ester hydrolase (CEH) and cholesterol oxidase (CHO):

cholesterol ester $\xrightarrow{\text{CEH}}$ cholesterol + fatty acid

cholesterol $\xrightarrow{\text{CHO}}$ cholestanone + H_2O_2

These two reactions are to take place in sequence, so that the amount of the final product, hydrogen peroxide, should reflect the sum of cholesterol ester and free cholesterol. The two enzymes, CEH and CHO, are used in the form of an immobilized enzyme column as shown in Fig. 5. Three different methods for utilizing two enzymes in one column were compared: (i) Sequential use of CEH and CHO, both immobilized, in this order, (ii) a mixed-bed use of the two immobilized enzymes and (iii) use of co-immobilized enzymes which means that CEH and CHO were simultaneously immobilized onto the same batch of glass beads. As shown in Fig. 6, only the co-immobilized enzyme reactor gave satisfactorily accurate results [14]. The interpretation is that the presence of CHO in the near vicinity of CEH is required for the rapid access by CEH of

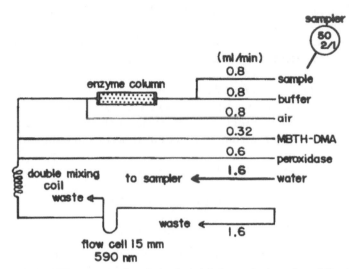

Figure 5. A segmented flow system of analysis of total cholesterol using coimmobilized cholesterol ester hydrolase and cholesterol oxidase column reactor integrated into Technicon AutoAnalyzer I. Taken from ref. 14.

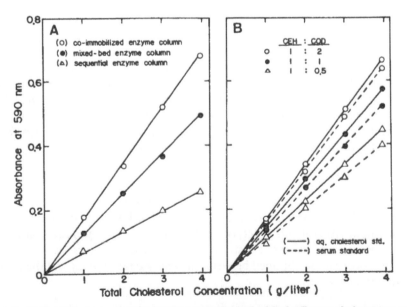

Figure 6. Calibration curves for total cholesterol obtained with the flow analytic system using immobilized cholesterol ester hydrolase (CEH) and cholesterol oxidase (COD) in the form of a column reactor. In (A): (O) co-immobilized enzyme; (●) mixed-bed enzyme column; (△) sequential enzyme column. The enzymatic activity ratio, in units, of CEH to COD was 1:2 in (A). In (B), the ratios were: (O), 1:2; (●), 1:1; (△), 1:0.5. Changes in absorbance were measured using aqueous cholesterol standard (——) or a pooled human serum standard (- - -). Taken from ref. 14.

esterified cholesterol in lipid miscells [14]. A similarly favorable effect of co-immobilization of the two enzymes which actually catalyze the two reaction in sequence was also found in the case of combined use of glucose oxidase and perioxidase for determining serum glucose [15].

4.2. Sequential use of immobilized enzymes

Contrary to the above-mentioned cases, the use of sequentially aligned immobilized enzymes is absolutely necessary in some other cases. For example, when creatinine in the serum is determined using immobilized creatinine deiminase (CD, creatinine iminohydrolase), the product to be measured is ammonia liberated from the substrate creatinine:

$$\text{creatinine} + H_2O \xrightarrow{\text{CD}} N\text{-methyhydantoin} + NH_3$$

$$NH_3 + \text{phenol} \xrightarrow{\text{hypochlorite}} \text{indophenol (color)}.$$

Therefore, the ammonia which already exists in the serum should be removed before the analyte comes into contact with the CD in the column reactor. This can be achieved by placing a pre-piece column of immobilized glutamate dehydrogenase (GLDH), which catalyzes the reaction:

$$NH_3 + \alpha\text{-oxoglutarate} + NADPH + H^+ \xrightarrow{\text{GLDH}} \text{L-glutamate}$$
$$+ NADP^+ + H_2O.$$

The whole system permits direct measurement on serum specimens at a feeding speed of 60 samples per hour [16]. If one mixes both CD and GLDH, whether free or immobilized, accurate measurement cannot be made. The alignment of two different immobilized enzymes in the correct order in one column permits the two reactions to take place in sequence which cannot be achieved otherwise.

A novel system of analysis has recently been devised for the sequential use of glutamate dehydrogenase (GLDH) and glutamate oxidase (GLXD), whereby ammonia is first convented into L-glutamate (as shown above), and then L-glutamate formed is oxidized to produce hydrogen peroxide.

$$\text{L-glutamate} + H_2O + O_2 \xrightarrow{\text{GLXD}} NH_3 + \alpha\text{-oxoglutarate} + H_2O_2$$

Hydrogen peroxide can be determined by any one of the commonly used methods [17]. It is essential that the GLDH and GLXD reactions are made to proceed in sequence in this order; otherwise one of the final products, ammonia, could be utilized as the substrate for the GLDH reaction so that no stoichiometry can be obtained between the analyte (ammonia) and the product (hydrogen peroxide).

5. BIOREACTORS FOR DIAGNOSTIC USE

5.1. Designing a new type of instrument

On the basis of accumulated knowledge and experience on how to use immobilized enzymes for clinical analysis, we decided to design an analytic instrument for use in routine clinical laboratories. For such a design, we chose the following three lines as innovative principles:

(i) to utilize a very rapid method of analysis such as the flow infection analysis (FIA),

(ii) to adopt a highly sensitive method of detection such as the chemiluminometry, and

(iii) to minimize the sample volume, which can in turn miniaturize the instrument as a whole.

The choice of FIA was based on our previous experience on the combined use of two or more enzymes [14–17]. Although the 'membrane type' use of immobilized enzymes, mostly in combination with an electrode, is prevailing, multi-step enzymatic analysis may not always be possible with membrane-embedded enzymes. On the contrary, a 'reactor-type' use (Fig. 1) with uni-directional flow of solution through it enables both the combined and sequential use of several enzymes depending upon whichever method is necessary. The use of an immobilized enzyme column together with FIA was thought to be ideal for our purpose.

The choice of chemiluminometry came from our observation that a large number of analytes in the serum and urine find specific oxidases in nature which produce hydrogen peroxide as the final product (Table 2). The well-known chemiluminescence emission that takes place when hydrogen peroxide is admixed with a mixture of luminol and potassium ferricyanide permits the determination of hydrogen peroxide at the 10^{-9} M level [18].

Scheme 1.

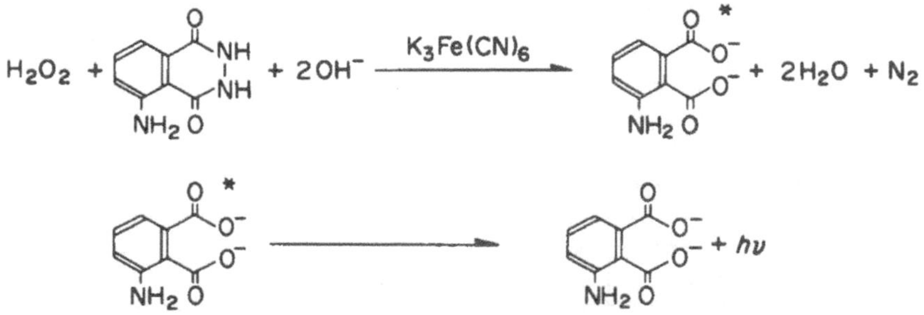

Figure 7 is an example showing how to determine 0.01–2.5 nmol glucose in $1\,\mu$l or less volume of sample by an FIA equipped with an immobilized glucose oxidase column, which gives analytic results within 10 s [19].

5.2. *Common final signal*

Since we had established a method for the determination of ammonia whereby the final signal emitted from the enzyme reactor is hydrogen peroxide [17], we could soon apply the same principle to the determination of urea [20]. Figure 8 illustrates the design of an enzyme column having three immobilized enzymes:

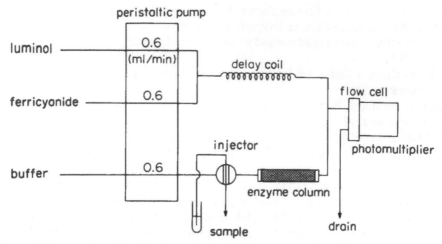

Figure 7. Diagram for a flow injection analysis of glucose using an immobilized glucose oxidase column reactor and chemiluminescence. Taken from ref. 18.

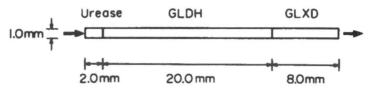

Figure 8. Design of the immobilized enzyme reactor for the chemiluminometric determination of urea in serum. GLDH, glutamate dehydrogenase; GLXD, glutamate oxidase.

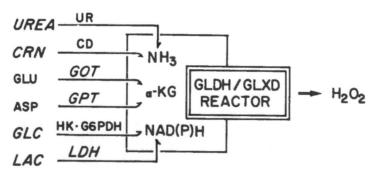

Figure 9. Versatile applicability of glutamate dehydrogenase (GLDH)/glutamate oxidase (GLXD) bioreactor. The bioreactor unit (shown as boxed) emits a common final signal, hydrogen peroxide, from a variety of analytes, of which items popular in clinical laboratories are shown as italicized. UR, urease; CRN, creatinine; CD, creatinine deiminase; GLU, glutamic acid; Asp, aspartic acid; GOT, aspartate transaminase; GPT, alanine transaminase; GLC, glucose; HK, hexokinase; LAC, lactate; LDH, lactate dehydrogenase; α-KG, α-oxoglutarate.

urease (UR), glutamate dehydrogenase (GLDH) and glutamate oxidase (GLXD), in this order. This column can be mounted in parallel with other enzyme columns to make a system having a sample-injecting device at one end and a chemiluminescence detecting device at the other end. Similarly, the enzyme reactor consisting of sequentially aligned GLDH and GLXD can also be used to determine a large number of other constituents of the serum and urine as are shown in Fig. 9. By this way, the final signal from diverse analytic reactions can be made common, i.e., in the form of hydrogen peroxide.

5.3. *Oligo-channeled instrument*

For constructing a commercial model, we invented a novel system of solution delivery which we wish to call 'one-shot' FIA (Fig. 10) [20]. By employing one-shot FIA, whereby computer-programmed syringe movements supply the required solutions in minimal volumes only at the needed moments, wastes of costly reagents can be largely avoided. Figure 11 shows an instrument put on the market in 1986. Practically satisfactory results of analysis are being recorded with this instrument (Fig. 12) [20]. Although this instrument houses only four immobilized enzyme columns at present, it is hoped that several other items will be added shortly, adhering to the same principle and dimensions of the present instrument.

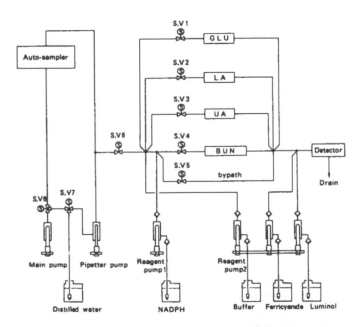

Figure 10. Flow diagram for 'one-shot flow injection analysis' with immobilized enzyme columns. Items of analysis are GLU for glucose, LA for lactate, UA for uric acid and BUN for blood ureanitrogen. S.V., solenoid value. Taken from ref. 19.

Figure 11. A commercial model of a diagnostic bioreactor (CL-760, Shimadzu Corporation, Kyoto, Japan). It measures 415 (H) × 585 (W) × 445 (D) mm, having a six-column capacity inside. The serum sample volume needed for simultaneous assays on four items is only 10 μl. It is operated at a speed of 56 samples/h. The immobilized enzymes in the reactor are stable for 4 weeks under daily use.

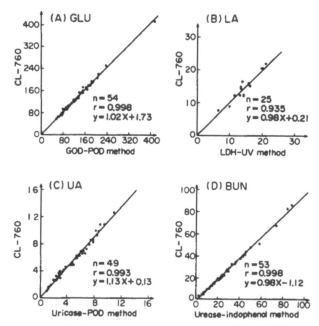

Figure 12. Correlation between the results of analysis obtained using a diagnostic bioreactor, Shimadzu CL-760, and those obtained by conventional methods as shown under the abscissae of the respective panels. The gratings of the figures are all in mg/dl. (A) glucose; (B) lactate; (C) uric acid; (D) blood urea-nitrogen. GOD, glucose oxidase; POD, peroxidase; LDH, lactate dehydrogenase; UV, ultraviolet absorption.

Acknowledgements

I wish to acknowledge excellent experimental works conducted and valuable discussions given by my colaborators, including Drs Jiro Endo, Masayoshi Tabata, Masayuki Totani and Hiroshi Nakano. This work was supported in parts by a grant-in-aid for special Project Research on Design of Multiphase Biomedical Materials and grants for Scientific Research from the Ministry of Education, Science and Culture, Japan, and grants for Life Science Project Research on Bioreactors from the Science and Technology Agency, Japan.

REFERENCES

1. Skeggs, L. T., An automatic method for colorimetric analysis. *Am. J. Clin. Pathol.* 28, 311–322 (1957).
2. Murachi, T., Use of immobilized enzyme reactors in automated clinical analysis. In: *Enzyme Engineering, vol. 6*, Eds Chibata, I., Fukui, S. and Wingard, L. B., Jr., Plenum Press, New York, 1982, pp. 369–375.
3. Murachi, T. and Tabata, M., Use of immobilized enzyme column reactors in clinical analysis. *Methods Enzymol.* 137, 260–271 (1988).
4. Tabata, M., Kido, T., Totani, M. and Murachi, T., Direct spectrophotometry of magnesium in serum after reaction with hexokinase and glucose-6-phosphate dehydrogenase. *Clin. Chem.* 31, 703–705 (1985).
5. Kido, T., Tabata, M., Totani, M. and Murachi, T., Enzymatic method for the determination of magnesium in urine using hexokinase or glucokinase and glucose-6-phosphate dehydrogenase, *Jpn. J. Clin. Chem.* 15, 146–151 (1986).
6. Tabata, M. and Murachi, T., Determination of inorganic phosphorus in serum using immobilized pyruvate kinase. *Biotechnol. Bioeng.* 25, 3013–3026 (1983).
7. Tabata, M., Kido, T., Ikemoto, M., Totani, M. and Murachi, T., Use of phospholipase D and choline oxidase for the enzymatic determination of calcium ion in serum. *J. Clin. Biochem. Nutr.* 1, 11–19 (1986).
8. Chibata, I. (Ed.), *Immobilized Enzymes. Research and Development*, Kodansha, Tokyo (1987).
9. Chiang, T. M. S. (Ed.), *Biochemical Applications of Immobilized Enzymes and Proteins, Vols 1 and 2*, Plenum Press, New York, 1977.
10. Weetall, H. H., *Immobilized Enzymes, Antigens, Antibodies and Peptides*, Marcell Dekker, New York, 1975.
11. Zaborsky, O. R. and Ogletree, J., The immobilization of glucose oxidase via activation of its carbohydrate residues. *Biochem. Biophys. Res. Commun.* 61, 210–216 (1974).
12. Nakane, P. K. and Kawaoi, A., Peroxidase-labeled antibody. A new method of conjugation. *J. Histochem. Cytochem.* 22, 1084–1091 (1974).
13. Kondo, T., Kojima, H., Ohkura, K., Ikeda, S. and Ito, K., Trial of new vessel access type glucose sensor for implantable artificial pancreas *in vivo*. *Trans. Am. Soc. Artif. Intern. Organs* 27, 250–252 (1981).
14. Tabata, M., Endo, J. and Murachi, T., Automated analysis of total cholesterol in serum using coimmobilized cholesterol ester hydrolase and cholesterol oxidase. *J. Appl. Biochem.* 3, 84–92 (1981).
15. Murachi, T., Sakaguchi, Y., Tabata, M., Sugahara, M. and Endo, J., Application of immobilized enzymes to clinical analysis: Use of coimmobilized glucose oxidase and peroxidase in column form. *Biochimie* 62, 581–585 (1980).
16. Tabata, M., Kido, T., Totani, M. and Murachi, T., Automated assay of creatinine in serum as simplified by the use of immobilized enzymes, creatinine deiminase and glutamate dehydrogenase. *Anal. Biochem.* 134, 44–49 (1983).
17. Murachi, T. and Tabata, M., Use of a bioreactor consisting of sequentially aligned L-glutamate dehydrogenase and L-glutamate oxidase for the determination of ammonia by chemiluminescence. *Biotechnol. Appl. Biochem.* 9, 303–309 (1987).
18. Tabata, M., Fukunaga, C., Ohyabu, M. and Murachi, T., Highly sensitive flow injection analysis of glucose and uric acid in serum using an immobilized enzyme column and chemiluminescence. *J. Appl. Biochem.* 6, 251–258 (1984).

19. Ohyabu, M., Fujimura, M., Tanimizu, K., Okuno, Y., Tabata, M., Totani, M. and Murachi, T., 'One-shot' flow injection analysis with immobilized enzyme columns: Clinical applications, *Anal. Sci.* **3**, 277–278 (1987).
20. Tabata, M. and Murachi, T., A chemiluminometric method for the determination of urea in serum using a three-enzyme bioreactor. *J. Bioluminesc. Chemiluminesc.* **2**, 63–67 (1988).

Multiphase Biomedical Materials, pp. 153–166 (1989)
T. Tsuruta and A. Nakajima (Eds)

Chapter 10

Synthetic polymer–protein hybrids: modification of L-asparaginase with polyethylene glycol

Y. INADA,[1] A. MATSUSHIMA,[1] Y. SAITO,[2] T. YOSHIMOTO,[2] H. NISHIMURA[3] and H. WADA[4]

[1]*Department of Materials Science and Technology, Toin University of Yokohama, 1614 Kuroganecho, Midoriku, Yokohama 227, Japan*
[2]*Laboratory of Biological Chemistry, Tokyo Institute of Technology, Ookayama, Meguroku, Tokyo 152, Japan*
[3]*Department of Pathology, School of Medicine, Juntendo University, Hongo, Bunkyoku, Tokyo 113, Japan*
[4]*Department of Pharmacology II, Osaka University, School of Medicine, Nakanoshima, Kitaku, Osaka 530, Japan*

Summary—Proteins can be modified by chemical binding of synthetic macromolecules to their surface. This modification can counteract some of the drawbacks of native proteins, and changes in their properties could be important for their use as protein drugs. One of the main advantages of construction of such modified proteins is the reduction of the immunoreactivity or immunogenicity of antigenically active proteins and another is the reduction in the clearance rate of protein drugs. For clinical use in the therapy of human leukemia, L-asparaginase from *Escherichia coli* was modified with polyethylene glycol. Administration of the modified asparaginase to three patients was therapeutically effective without causing any allergic reaction.

1. INTRODUCTION

Chemical modification of proteins with various kinds of modifiers including polyethylene glycol which bind to specific amino acids has been used for two main purposes: (i) for analysis of protein structures such as the active sites of enzymes and the states of amino acid residues [1, 2] and (ii) for altering and improving the native functions of proteins and endowing them with new functions. For example, synthetic polymers such as polyethylene glycol (PEG) have been used to alter the immunoreactivity or immunogenicity of antigenic proteins [3–7] or to make enzymes soluble and active in organic solvents with a view to their use as new bioreactors [8, 9].

L-Asparaginase from *Escherichia coli* has been used clinically for therapy of leukemia and lymphosarcoma. Capizzi [10] demonstrated that administration of asparaginase-methotrexate was therapeutically effective in patients with acute lymphoblastic leukemia. The disadvantage of this therapy lies in the immunological side effects of treatment, ranging in severity from mild allergic reactions to anaphylactic shock, as asparaginase is a foreign protein for humans. To overcome this drawback, we modified asparaginase with poly-ethylene glycol.

This paper describes the synthesis of L-asparaginase chemically modified with polyethylene glycol, the physicochemical properties of the modified protein and a study on its therapeutic effect.

2. SYNTHESIS OF POLYETHYLENE GLYCOL-MODIFIED L-ASPARAGINASE

One of the best ways of modifying L-asparaginase is the substitution method using cyanuric chloride (2,4,6-trichloro-s-triazine) and polyethylene glycol (PEG) [5]. As shown in Fig. 1, when monomethoxypolyethylene glycol with a molecular weight of approximately 5000 is treated with cyanuric chloride, two of the three chlorine atoms in the cyanuric chloride molecule are replaced by PEG with the formation of 2,4-bis(O-methoxypolyethylene glycol)-6-chloro-s-triazine, named activated PEG_2. The one remaining chlorine atom in activated PEG_2 can then combine with either the terminal amino groups and/or lysine ε-amino groups of asparaginase. For modification of asparaginase, activated PEG_2 is more effective than activated PEG_1 (2-O-methoxypolyethylene glycol-4,6-dichloro-s-triazine), in which only one of the three chlorine atoms in the cyanuric chloride molecule is replaced by PEG [11]. Figure 2 shows the homogeneity of activated PEG_2 [12]. The peak of activated PEG_2 synthesized (curve A) corresponded to a molecular weight of 10 000 and was separated from the peak of activated PEG_1 with a molecular weight of 5000 (curve B) and the peak of cyanuric acid formed by the hydrolysis of cyanuric chloride (curve C). The coupling reaction should be carried out under mild conditions (at room

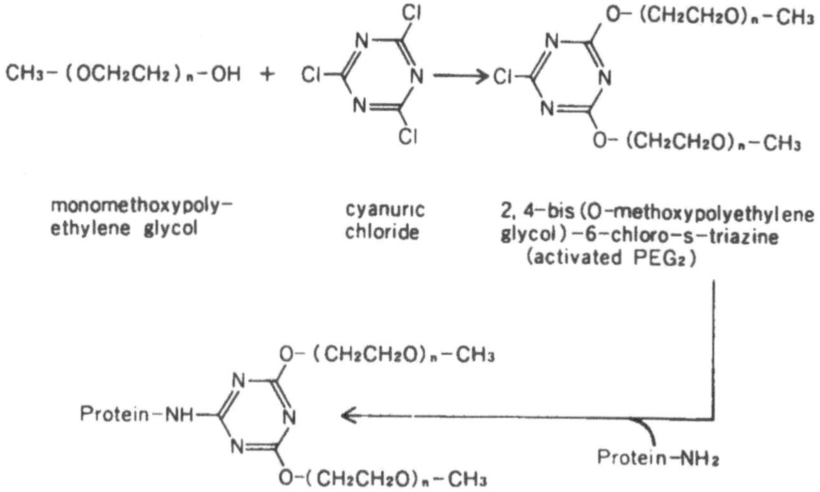

Figure 1. Synthesis of PEG_2-modified L-asparaginase.

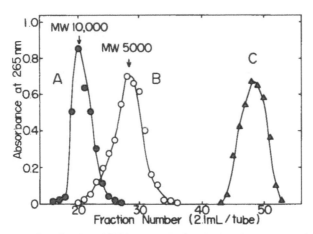

Figure 2. Homogeneity of activated PEG$_2$. A sephadex G-100 column was used for gel filtration chromatography. Curves A, B and C: activated PEG$_2$, activated PEG$_1$ and cyanuric acid, respectively.

temperature and neutral pH and in aqueous solution) to avoid protein denaturation. The degree of modification can be controlled by varying the molar ratio of the modifier to the protein in the reaction system.

3. CHARACTERIZATION OF PEG$_2$-MODIFIED L-ASPARAGINASE

L-Asparaginase from *E. coli* has a molecular weight of approximately 136 000 and is composed of four identical subunits. Its primary structure was clarified by Maita *et al.* [13]. It contains a total of 92 amino groups (four terminal amino groups and 88 lysine ε-amino groups). The properties of the modified asparaginase hoped to be acquired by coupling with activated PEG$_2$ were (i) reduced immunoreactivity towards anti-asparaginase antibodies, (ii) reduced immunogenicity, (iii) a prolonged clearance time, (iv) improved antitumor activity and (v) retention of enzymic activity.

First, we examined the effect of the chain length of monomethoxypolyethylene glycol on the properties of PEG$_2$-modified asparaginase, using activated PEG$_2$s prepared from monomethoxypolyethylene glycols with molecular weights of 350, 750, 2000 and 5000 [12]. Table 1 shows the enzymic activities and immunoreactivities of the resultant PEG$_2$-modified asparaginases, each with a similar degree (29–35%) of modification. Modification of asparaginase with higher molecular weight forms of monomethoxypolyethylene glycol resulted in greater reduction of immunoreactivity with anti-asparaginase antibodies and higher retention of enzymic activity. These results may have been because lower molecular weight forms can penetrate into the inside of the asparaginase molecule and react with the inner amino groups, thus disrupting the active conformation and reducing the enzymic activity. These results showed that modification of asparaginase with monomethoxypolyethylene glycol of molecular weight 5000 resulted in almost complete loss of immunoreactivity but retention of high enzymic activity.

Table 1.
Effect of molecular weight of polyethylene glycol on enzymic activity and immunoreactivity of PEG$_2$-modified L-asparaginase

Average molecular weight of PEG[a]	Degree of modification (%)[b]	Enzymic activity (IU/mg of protein)	Immunoreactivity (%)[c]
(control)	0	250	100
350	29	23	11.0
750	29	43	9.0
2000	34	63	1.9
5000	35	73	0.9

[a] Monomethoxypolyethylene glycol.
[b] The asparaginase molecule contains a total of 92 amino groups.
[c] Percentage of the control value in the precipitin reaction.

The modified asparaginase showed the same pH- and temperature-dependency of activity as the native asparaginase [12]. The optimum pH was about 7.0 and the profile of pH stability was the same as that of the pH dependency. The optimum reaction temperature was about 50°C, with rapid decrease in enzymic activity above 65°C and below 40°C. Native and modified asparaginase are both stable below 45°C. Kinetic studies on the modified asparaginase showed that its Michaelis–Menten constant for L-asparagine, the K_m value and maximum rate, V_{max}, were 7 μM and 73 IU/mg of protein, respectively. This K_m value was the same as that of the native enzyme, while the V_{max} was about one-third of that (250 IU/mg of protein) of native asparaginase. The native asparaginase showed very low activity with L-glutamine (V_{max} and K_m value, 7% and approximately 300 times those with L-asparagine, respectively). The modified enzyme also retained the high substrate specificity.

4. REDUCTION OF IMMUNOGENICITY AND PROLONGATION OF CLEARANCE TIME

Native and the modified asparaginase were injected intravenously into mice to test their immunogenicity [14]. The average antibody titers 7 days after the second and the third injections of the native asparaginase were 2^4 and 2^7, respectively. On the other hand, scarcely any antibody response was seen 7 days after the second immunization with PEG$_2$-modified asparaginase, and only a low titer (2^2–2^3) of antibody was observed after the third immunization (Fig. 3).

The modified asparaginase was supposed to be much less susceptible to digestion by trypsin than native asparaginase [3]. Digestion with trypsin for 30 min caused significant loss of activity of native asparaginase (80%), but only 10% loss of activity of the modified asparaginase. The high resistance of PEG$_2$-modified asparaginase to trypsin may be due to modification of amino groups in lysine residues, which are sites of cleavage by trypsin.

The *in vivo* effects of single intraperitoneal injections of the native and modified asparaginase into rats were studied. Concomitantly, their clearance rates from the blood were estimated [14]. Figure 4 shows the changes with time

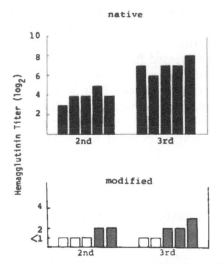

Figure 3. Immunogenicities of native (upper columns) and PEG$_2$-modified (lower columns) L-asparaginases. Mice were immunized repeatedly with 20 μg of enzyme protein and bled 7 days after the second or third immunization. Columns show antibody titers of the sera of individual mice. Solid and open columns represent significant titers (more that 2^2) and insignificant titers, respectively.

in the asparaginase activity and asparagine level in the sera of rats after administrations of native and the modified asparaginases. When native asparaginase (80 IU/kg) was injected intraperitoneally, its activity soon appeared in the serum and reached a peak after about 3 h. Subsequently, the activity rapidly decreased, becoming undetectable (0.001 IU/ml of serum) within 41 h. Asparagine in the serum decreased to less than 5 μm within 30 min after the

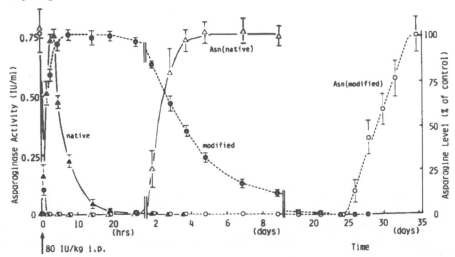

Figure 4. L-Asparaginase activity and L-asparagine concentration in rat sera after intraperitoneal injection of native or PEG$_2$-modified L-asparaginase. Asparaginase activity (closed symbols): ▲, native enzyme; ●, modified enzyme. Asparagine concentration (open symbols): △, with native enzyme; ○, with modified enzyme.

injection and returned to the normal level after 3 days. After injection of modified asparaginase, its activity was detected in the serum within 30 min, reached a maximal level after about 8 h and remained at the peak level for about 20 h. Then the activity decreased far more slowly than that of native asparaginase, and so was still detectable after 3 weeks. Correspondingly, the asparagine level in the serum decreased within 30 min and was still undetectable 3 weeks later. This depletion of asparagine persisted for more than 20 times longer than that after injection of native asparaginase. The half-lives of the native and modified asparaginases in the serum were calculated to be 2.9 and 56 h, respectively. The increased circulatory life-time of modified asparaginase may be due to its slow elimination because of its large molecular size and resistance to proteinase endowed by PEG_2-covered amino groups.

5. INCREASED ANTITUMOR ACTIVITY

An increased antitumor activity of the modified asparaginase *in vivo* could be expected from its half-life and the persistence of high enzymic activity in the serum after its injection. Gardner lymphoma cells (10^6), which are known to be L-asparagine dependent, were implanted into mice and 1 day later the animals were treated with native or modified asparaginase [15]. The survival times of mice treated with the modified asparaginase were significantly more than those of mice treated with the same dose of native asparaginase (Table 2). Three of

Table 2.
Comparison of therapeutic efficacies of native and PEG_2-modified L-asparaginase against Gardner lymphoma (6C3HED)[a]

Enzyme	Dose (IU/Mouse)	Survival time (days × no. of mice)	Mean (days)	P[b]
control	—	12 × 4, 13 × 3, 14 × 2, 15	13.0	
native	0.5	11, 14, 15, 17, 18	15.0	>0.05[d]
L-asparaginase	0.5 × 3[c]	18 × 4, 19	18.2	<0.01[d]
	1.0	14, 15, 17, 19 × 2	16.8	<0.01[d]
	2.0	16, 17, 19, 22, 23	19.4	<0.01[d]
	4.0	17, 19 × 2, 23, 26	20.8	<0.01[d]
	8.0	19 × 2, 22, 26, 60+[e]	29.2+	<0.01[d]
modified	0.5	18 × 3, 19 × 2	18.4	<0.025[f]
L-asparaginase	1.0	17 × 2, 20 × 3	18.8	<0.05[f]
	2.0	20 × 2, 23, 26, 27	23.2	<0.05[f]
	4.0	20, 23, 26 × 2, 31	25.2	<0.05[f]
	8.0	26 × 2, 60+[e] × 3	46.4+	<0.05[f]

[a] Normal CBA mice (5–8 weeks old) were inoculated i.p. with 10^6 lymphoma cells on day 0 and treated with the indicated dose of enzyme on day 1. The survival times of the mice after tumor implantation were measured. Groups of five mice were used except as controls (10 mice).
[b] Calculated by the Wilcoxon rank sum test.
[c] These mice were given three i.p. injections of native asparaginase (0.5 IU/mouse × 3) on days 1, 3 and 5.
[d] Compared with the survival time of the control group.
[e] The experiment was terminated on day 60. All surviving mice on day 60 were tumor-free.
[f] Compared with the survival time of the group treated with the corresponding dose of native enzyme.

Table 3.
Therapeutic efficacy of PEG$_2$-modified L-asparaginase against Gardner lymphoma (6C3HED) in preimmune mice[a]

Enzyme	Survival time (days × no. of mice)	Mean (days)	P[b]
control	12 × 2, 13 × 2, 15	13.0	
native L-asparaginase	12, 13 × 3, 14	13.0	>0.05[c]
modified L-asparaginase	25, 27, 31, 60 +[d] × 2	40.6 +	<0.01[c]

[a] CBA mice were preimmunized by two i.p. injections of 20 μg of native enzyme with Freund's complete adjuvant with an interval of 1 week between injections. One week after the second immunization, the mice were inoculated i.p. with 10^6 lymphoma cells on day 0 and treated with 8.0 IU/mouse of each enzyme on day 1. Groups of five mice were used.
[b] Calculated by the Wilcoxon rank sum test.
[c] Compared with the survival time of the control group.
[d] The experiment was terminated on day 60. All surviving mice on day 60 were tumor-free.

five mice treated with 8.0 IU/mouse of PEG$_2$-modified asparaginase were still alive on day 60 and were found to be tumor-free. The survival of mice treated 3 times with native enzyme (0.5 IU/mouse × 3 on days 1, 3 and 5) was the same as that of mice treated once with the modified asparaginase (0.5 IU/mouse on day 1). Thus, the therapeutic activity of PEG$_2$-modified asparaginase appeared to be superior to that of the native enzyme. Table 3 shows the effect of pre-immunization with native asparaginase on the antitumor activities of native and modified asparaginase [15]. Treatment with native asparaginase (8.0 IU/mouse) did not increase the survival time of preimmune mice, whereas two out of five mice treated with the modified asparaginase (8.0 IU/mouse), were alive on day 60 and were tumor-free. Thus preimmunization with native asparaginase caused specific antibody formation in mice and deprived native asparaginase of antitumor activity, but did not affect the therapeutic activity of modified asparaginase. These results indicate that anti-asparaginase antibodies do not bind to modified asparaginase.

As a preclinical trial, a therapeutic study was performed with native and modified asparaginases in a dog with spontaneous lymphosarcoma. This lymphosarcoma had been diagnosed by histological examination of an aspiration biopsy specimen of a neck lymph node 3 days before the trial. Biochemical tests were performed before and after the intravenous injection of native asparaginase (500 IU) followed by modified asparaginase (50 IU) into the dog (14 kg) [12]. The results are shown in Fig. 5 and Table 4. Asparaginase activity appeared in the serum after injection of native asparaginase and decreased with a half-life of about 1 day. During this time, the size of the neck lymph nodes and the volume of ascites decreased, indicating that the tumor was asparaginase-sensitive. During the 12 days following the first injection of the native enzyme, the lymph nodes first shrank and then increased in size again, due to a complete disappearance of the enzymic activity and restoration of the asparagine level in the serum. Injection of the modified asparaginase at one-tenth the dose of enzymic activity of the native enzyme in the first injection, however,

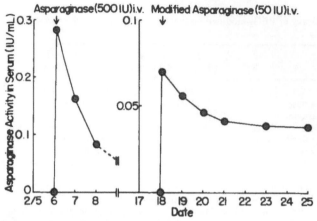

Figure 5. Prolonged clearance time of PEG_2-modified L-asparaginase in a lymphosarcoma-bearing dog. The dog was treated intravenously with native asparaginase (500 IU) followed by PEG_2-modified asparaginase (50 IU). On the dated indicated, blood was drawn from a peripheral vein for measurement of enzymic activity.

caused striking shrinkage of the neck lymphadenoma and reduction of the ascites, due to complete depletion of serum asparagine. The level of glutamine, however, was not decreased (in fact it was slightly increased), due to the high specificity of the enzyme. The lymph nodes were no longer palpable 5 days after the second injection. The clearance rate of the modified enzyme from the blood was much slower than that of the native enzyme. The same biochemical and pharmacokinetic results were obtained in similar trials in dogs without

Table 4.
Therapeutic effectiveness of PEG_2-modified L-asparaginase on a lymphosarcoma-bearing dog

			Native L-asparaginase ↓ (500 IU) i.v.				Modified L-asparaginase ↓ (50 IU) i.v.			
Date	Feb.	5	6	7	8	17	18	19	23	25
Amino acid	Asp		21	92			10	35	59	69
levels in	Asn		75	n.d.			73	n.d.	n.d.	n.d.
serum (μM)	Glu		158	336			82	181	203	346
	Gln		796	1211			574	633	866	977
Peripheral	RBC	238×10^4				280×10^4				345×10^4
blood (/mm³)	WBC	26×10^3				20×10^3				11×10^3
Ascites		+++			+	+++	–	–	–
lymph nodes		4×4			2×2	3×4		2×3	0×0	0×0
(cm × cm)										

The dog was treated intravenously with native asparaginase (500 IU) followed by modified asparaginase (50 IU). On the dates indicated, blood was drawn from a peripheral vein to determine the serum amino acid levels and the numbers of red blood cells (RBC) and white blood cells (WBC). The values of amino acids in the serum shown on Feb. 6th and 18th were obtained before administration of native and modified asparaginase, respectively. Asp, aspartic acid; Asn, asparagine; Glu, glutamic acid; Gln, glutamine; n.d., not detectable (less that 0.5 μM).

lymphosarcoma. During these trials, no clinical or histological evidence of toxicity of the modified asparaginase was observed. Biopsy on day 8 after the injection of modified asparaginase showed no malignant cells in the neck lymph nodes.

6. CLINICAL APPLICATION TO CHILDREN WITH ACUTE LEUKEMIA

The modified asparaginase was applied to three children with leukemia who developed an allergic reaction to the native enzyme during methotrexate-asparaginase sequential maintenance therapy [12].

⟨Case 1⟩ W.K. (born 9/10/1972, male) was diagnosed as having acute lymphocytic leukemia (ALL) at the age of 6 and recovered until he suffered testicular and meningeal relapse at the age of 9. His first hematological relapse occurred at the age of 11. After remission-induction therapy, he was placed on a sequential methotrexate-asparaginase maintenance regimen. He had received 150 000 IU of asparaginase for induction therapy and 20 000 IU for metho-trexate-asparaginase sequential therapy by the time of the second course of methotrexate-asparaginase, when he suffered anaphylactic shock. Anaphylac-tic shock occurred with hypotension, severe coughing, vomiting, abdominal pain and urticaria 30 min after the second course of asparaginase. These symp-toms were relieved by injection of antihistamine and hydrocortisone, but the last two symptoms lasted for about 12 h. Because of this anaphylactic reaction, the maintenance regimen with native asparaginase could not be continued. Therefore, we resumed the maintenance regimen using the modified aspara-ginase with prednisolone in an attempt to avoid the anaphylactic reaction. No allergic symptoms were seen. The serum asparagine, which dropped from 12 μM to an undetectable level after administration of the modified enzyme, remained undetectable for more than 2 weeks. Prednisolone was omitted in the following course of the regimen. He has now received more than 10 courses of this regimen (in 5 months) without any allergic reaction and no side effects such as liver dysfunction, coagulation disorder or diabetes mellitus attributable to aspara-ginase have been noted.

⟨Case 2⟩ A.S. (born 4/23/1971, female) was diagnosed as having ALL at the age of 10 and was recovering until a first hematological relapse occurred at the age of 13. After reinduction therapy, she was placed on the maintenance regimen. She received 225 000 IU of asparaginase for remission induction therapy and 14 000 IU for maintenance therapy up to the time of the 14th course of asparaginase, when she suffered from a severe coughing, abdominal pain and urticaria. Modified asparaginase was then used in place of native asparaginase. No allergic reactions occurred and her serum asparagine level dropped from 72 μM to an undetectable level which persisted for 2 weeks. She has now received more than 10 courses of this regimen without any side effects.

⟨Case 3⟩ K.F. (born 5/12/1977, male) was diagnosed as having acute mono-cytic leukemia (AMoL) at the age of 4. After the second hematological relapse, he was placed on an acute non-lymphocytic leukemia (ANLL) maintenance regimen consisting of two courses of methotrexate-asparaginase and one course of continuous cytosine arabinoside for 5 days and a bolus injection of aclacino-mycin for 3 days every 2 weeks. He received 10 500 IU of asparaginase for maintenance therapy by the time of the 15th course of asparaginase, when he suffered from a severe cough, abdominal pain, vomiting and urticaria. From the next course, he was treated with the modified asparaginase and did not show any allergic reactions.

The half-life of modified asparaginase after an injection of 200 IU was in the order of days, while that of the native enzyme after a much higher dose of 10 000 IU was in the order of hours. This phenomenal difference may be attributable to the following factors. The modified enzyme is more resistant to various proteolytic enzymes such as trypsin and chymotrypsin, less potent in stimulating

Table 5.
Polyethylene glycol–protein hybrids

Protein	Use	Reference
To reduce antigenicity and prolong clearance time		
adenosine deaminase		[16]
arginase	anti-tumor	[17, 18]
asparaginase	anti-tumor	[3, 6, 12, 14, 15, 19]
batroxobin	anti-thrombotic	[5]
blood coagulation factor VIII, IX	replacement therapy for hemophilia	[20]
bovine serum albumin		[21]
catalase		[22]
elastase	therapy of arteriosclerosis	[23]
α-galactosidase	enzyme replacement therapy	[24]
β-glucosidase	enzyme replacement therapy	[24]
β-glucuronidase	enzyme replacement therapy	[25]
hemoglobin	artificial blood	[26–27]
immunoglobulin G	immunoglobulin therapy	[28]
insulin	therapy of diabetes mellitus	[29]
lactoferrin		[30]
α_2-macroglobulin		[30]
phenylalanine ammonia-lyase	anti-tumor	[31]
ragweed pollen	therapy of type I allergy	[32–34]
streptokinase	anti-thrombotic	[35–37]
superoxide dismutase	anti-inflammatory	[30, 38]
trypsin		[39]
tryptophanase	anti-tumor	[70]
uricase	therapy of hyperuricemia and gout	[4, 40–42]
urokinase	anti-thrombotic	[43, 44]
To make soluble and active in organic solvents		
catalase	decomposition of H_2O_2	[45]
chymotrypsin	formation of acid–amide linkage	[46]
papain		[47]
lipase	ester synthesis, ester exchange, 'magnetic enzyme'	[48–68]
peroxidase	oxidation with H_2O_2	[69]

the immune system and less easily taken up by reticuloendothelial cells than the native enzyme. We have been able to cut down the dose of asparaginase injected from 10 000 to 200 IU because of the prolonged clearance time of the enzyme resulting from its chemical modification. In fact, if we used the same dosage of the modified enzyme as that with the native enzyme (10 000 IU), we would expect that the excessively high asparaginase activity would hydrolyze the abundant glutamine in the blood with abrupt production of ammonia.

Repeated administration of the modified asparaginase did not induce immunological side effects in patients who had already showed anaphylactic reactions to native enzyme. Thus it can be concluded that PEG_2-modified L-asparaginase is not immunoreactive with anti-asparaginase antibody or immunogenic, producing its own antibody, and that it is therapeutically useful against leukemia. This success opens up the possibility of developing new types of enzyme drugs in the future.

7. POLYETHYLENE GLYCOL-PROTEIN HYBRIDS

The present article deals with a hybridized drug composed of *E. coli* L-asparaginase and polyethylene glycol (PEG) which is a non-immunogenic synthetic polymer. There have been many similar lines of work on hybrid molecules, as summarized in Table 5. In this way, many kinds of hybrid compounds of proteins, biologically active substances, synthetic polymers and inorganic materials can be created for uses in the fields of medicine, pharmacy, technology and agriculture.

Acknowledgements
We are deeply indebted to Dr. Yoshinori Kamisaki, School of Medicine, Tottori University, Dr Masahiro Sako and Dr Giichi Tsujino, Children's Medical Center of Osaka City, Professor Tetsuo Taguchi, Research Institute for Microbial Diseases, Osaka University and Professor Schusaku Noda, Faculty of Agriculture, University of Osaka Prefecture, for their cooperation.

REFERENCES

1. Vallee, B. L. and Riordan, J. F., Chemical approaches to the properties of active sites of enzymes. *Annu. Rev. Biochem.* **38**, 733–794 (1969).
2. Means, G. E. and Feeney, R. E., *Chemical Modification of Proteins*. Holden-Day, Inc., San Francisco (1971).
3. Matsushima, A., Nishimura, H., Ashihara, Y., Yokota, Y. and Inada, Y., Modification of *E. coli* asparaginase with 2,4-bis(*O*-methoxypolyethylene glycol)-6-chloro-s-triazine (activated PEG_2): Disappearance of binding ability towards anti-serum and retention of enzymic activity. *Chem. Lett.* **7**, 773–776 (1980).
4. Nishimura, H., Matsushima, A. and Inada, Y., Improved modification of yeast uricase with polyethylene glycol, accompanied with non-immunoreactivity towards anti-uricase serum and high enzymic activity. *Enzyme* **26**, 49–53 (1981).
5. Nishimura, H., Takahashi, K., Sakurai, K., Fujinuma, K., Imamura, Y., Ooba, M. and Inada, Y., Modification of batroxobin with activated polyethylene glycol: Reduction of binding ability towards anti-batroxobin antibody and retention of defibrinogenation activity in circulation of preimmunized dogs. *Life Sci.* **33**, 1467–1473 (1983).
6. Kawamura, K., Igarashi, T., Fujii, T., Kamisaki, Y., Wada, H. and Kishimoto, S., Immune responses to polyethylene glycol modified L-asparaginase in mice. *Int. Archs Allergy Appl. Immunol.* **76**, 324–330 (1985).

7. Inada, Y., Yoshimoto, T., Matsushima, A. and Saito, Y., Engineering physicochemical and biological properties of proteins by chemical modification. *Trends Biotechnol.* **4**, 68–73 (1986).

8. Inada, Y., Takahashi, K., Yoshimoto, T., Ajima, A., Matsushima, A. and Saito, Y., Application of polyethylene glycol-modified enzymes in biotechnological processes: Organic solvent-soluble enzymes. *Trends Biotechnol.* **4**, 190–194 (1986).

9. Inada, Y., Takahashi, K., Yoshimoto, T., Kodera, Y., Matsushima, A. and Saito, Y., Application of PEG-enzyme and magnetite-PEG-enzyme conjugates for biotechnological processes. *Trends Biotechnol.* **6**, 131–134 (1988).

10. Capizzi, R. L., Asparaginase-methotrexate in combination chemotherapy: Schedule-dependent differential effects on normal versus neoplastic cells. *Cancer Treat. Rep.* **65**, (Suppl. 4), 115–121 (1981).

11. Park, Y. K., Abuchowski, A., Davis, S. and Davis, F., Pharmacology of *Escherichia coli*-L-asparaginase polyethylene glycol adduct. *Anticancer Res.* **1**, 373–376 (1981).

12. Yoshimoto, T., Nishimura, H., Saito, Y., Sakurai, K., Kamisaki, Y., Wada, H., Sako, M., Tsujino, G. and Inada, Y., Characterization of polyethylene glycol-modified L-asparaginase from *Escherichia coli* and its application to therapy of leukemia. *Jpn. J. Cancer Res. (Gann)* **77**, 1264–1270 (1986).

13. Maita, T., Morokuma, K. and Matsuda, G., Amino acid sequence of L-asparaginase from *Escherichia coli*. *J. Biochem.* **76**, 1351–1354 (1974).

14. Kamisaki, Y., Wada, H., Yagura, T., Matsushima, A. and Inada, Y., Reduction in immunogenicity and clearance rate of *Escherichia coli* L-asparaginase by modification with monomethoxypolyethylene glycol. *J. Pharmacol. Exp. Ther.* **216**, 410–414 (1981).

15. Kamisaki, Y., Wada, H., Yagura, T., Nishimura, H., Matsushima, A. and Inada, Y., Increased antitumor activity of *Escherichia coli* L-asparaginase by modification with mono-methoxypolyethylene glycol. *Jpn. J. Cancer Res. (Gann)* **73**, 470–474 (1982).

16. Steinbeck, W., Covalent attachment of polyethylene glycol to adenosine deaminase and its effect on blood circulating life and immunogenicity of the conjugate. Honors Thesis, Rutgers University (1978).

17. Savoca, K. V., Abuchowski, A., van Es, T., Davis, F. F. and Palczuk, N. C., Preparation of a non-immunogenic arginase by the covalent attachment of polyethylene glycol. *Biochim. Biophys. Acta* **578**, 47–53 (1979).

18. Savoca, K. V., Davis, F. F., van Es, T., McCoy, J. R. and Palczuk, N. C., Cancer therapy with chemically modified enzymes. II. The therapeutic effectiveness of arginase, and arginase modified by the covalent attachment of polyethylene glycol, on the taper liver tumor and the L5178Y murine leukemia. *Cancer Biochem. Biophys.* **7**, 261–268 (1984).

19. Abuchowski, A., Kazo, G. M., Verhoest, Jr., C. R., van Es, T., Kafkewitz, D., Nucci, M. L., Viau, A. T. and Davis, F. F., Cancer therapy with chemically modified enzymes. I. Antitumor properties of polyethylene glycol-asparaginase conjugates. *Cancer Biochem. Biophys.* **7**, 175–186 (1984).

20. Sakuragawa, N., Kondo, S., Kondo, K. and Niwa, M., Oral administration of factor VIII or IX concentrates preparation using dogs. *Acta Me. Biol. (Niigata)*, **34**, 77–83 (1986).

21. Abuchowski, A., van Es, T., Palczuk, N. C. and Davis, F. F., Alteration of immunological properties of bovine serum albumin by covalent attachment of polyethylene glycol. *J. Biol. Chem.* **252**, 3578–3581 (1977).

22. Abuchowski, A., McCoy, J. R., Palczuk, N. C., van Es, T. and Davis, F. F., Effect of covalent attachment of polyethylene glycol on immunogenicity and circulating life of bovine liver catalase. *J. Biol. Chem.* **252**, 3582–3586 (1977).

23. Koide, A. and Kobayashi, S., Modification of amino groups in porcine pancreatic elastase with polyethylene glycol in relation to binding ability towards anti-serum and to enzymic activity. *Biochem. Biophys. Res. Commun.* **111**, 659–667 (1983).

24. Wieder, K. J. and Davis, F. F., Enzyme therapy II. Effect of covalent attachment of polyethylene glycol on biochemical parameters and immunological determinants of β-glucosidase and α-galactosidase. *J. Appl. Biochem.* **5**, 337–347 (1983).

25. Lisi, P. J., van Es, T., Abuchowski, A., Palczuk, N. C. and Davis, F. F., Enzyme therapy I. Polyethylene glycol: β-glucuronidase conjugates as potential therapeutic agents in acid mucopolysaccharidosis. *J. Appl. Biochem.* **4**, 19–33 (1982).

26. Ajisaka, K. and Iwashita, Y., Modification of human hemoglobin with polyethylene glycol: A new candidate for blood substitute. *Biochem. Biophys. Res. Commun.* **97**, 1076–1081 (1980).

27. Leonard, M. and Dellacherie, E., Acylation of human hemoglobin with polyoxyethylene derivatives. *Biochim. Biophys. Acta* **791**, 219–225 (1984).

28. Suzuki, T., Kanbara, N., Tomono, T., Hayashi, N. and Shinohara, I., Physicochemical and

biological properties of poly(ethylene glycol)-coupled immunoglobulin. *G. Biochim. Biophys. Acta* **788**, 248–255 (1984).

29. Ehrat, M. and Luisi, P. L., Synthesis and spectroscopic characterization of insulin derivatives containing one or two poly(ethylene oxide) chains at specific positions. *Biopolymer* **22**, 569–573 (1983).
30. Beauchamp, C. O., Gonias, S. L., Menapace, D. P. and Pizzo, S. V., A new procedure for the synthesis of polyethylene glycol-protein adducts; Effects on function, receptor recognition, and clearance of superoxide dismutase, lactoferrin, and α_2-macroglobulin. *Anal. Biochem.* **131**, 25–33 (1983).
31. Wieder, K. J., Palczuk, N. C., van Es, T. and Davis, F. F., Some properties of polyethylene glycol: phenylalanine ammonia-lyase adducts. *J. Biol. Chem.* **254**, 12579–12587 (1979).
32. Lee, W. Y. and Sehon, A. H., Abrogation of reaginic antibodies with modified allergens. *Nature* **267**, 618–619 (1977).
33. Lee, W. Y. and Sehon, A. H., Suppression of reaginic antibodies with modified allergens I. Reduction in allergenicity of protein allergens by conjugation to polyethylene glycol. *Int. Archs Allergy Appl. Immunol.* **56**, 193–206 (1978).
34. Lee, W. Y. and Sehon, A. H., Suppression of reaginic antibodies with modified allergens II. Abrogation of reaginic antibodies with allergens conjugated to polyethylene glycol. *Int. Archs Allergy Appl. Immunol.* **56**, 159–170 (1978).
35. Koide, A., Suzuki, S. and Kobayashi, S., Preparation of polyethylene glycol-modified streptokinase with disappearance of binding ability towards anti-serum and retention of activity. *FEBS Lett.* **143**, 73–76 (1982).
36. Newmark, J., Abuchowski, A. and Murano, G., Preparation and properties of adducts of streptokinase and streptokinase-plasmin complex with polyethylene glycol and pluronic polyol F38. *J. Appl. Biochem.* **4**, 185–189. (1982).
37. Rajagopalan, S., Gonias, S. L. and Pizzo, S. V., A nonantigenic covalent streptokinase-polyethylene glycol complex with plasminogen activator function. *J. Clin. Invest.* **75**, 413–419 (1985).
38. Pyatak, P. S., Abuchowski, A. and Davis, F. F., Preparation of a polyethylene glycol: super-oxide dismutase adduct, and an examination of its blood circulating life and anti-inflammatory activity. *Res. Commun. Chem. Pathol. Pharmacol.* **29**, 113–127 (1980).
39. Abuchowski, A. and Davis, F. F., Preparation and properties of polyethylene glycol–trypsin adducts. *Biochim. Biophys. Acta* **578**, 41–46 (1979).
40. Nishimura, H., Ashihara, Y., Matsushima, A. and Inada, Y., Modification of yeast uricase with polyethylene glycol: Disappearance of binding ability towards anti-uricase serum. *Enzyme* **24**, 261–264 (1979).
41. Chen, R. H.-L., Abuchowski, A., van Es, T., Palczuk, N. C. and Davis, F. F., Properties of two urate oxidases modified by the covalent attachment of poly(ethylene glycol). *Biochim. Biophys. Acta* **660**, 293–298 (1981).
42. Abuchowski, A., Karp, D. and Davis, F. F., Reduction of plasma urate levels in the cockerel with polyethylene glycol-uricase. *J. Pharmacol. Exp. Ther.* **219**, 352–354 (1981).
43. Igarashi, M., Takatsuka, J., Shiba, T., Takeuchi, S., Kinoshita, T., Nakahara, T. and Shimizu, K., Chemical modification of urokinase with polyethylene glycol and its *in vitro* and *in vivo* study. *Ketsueki To Myakkan (Blood and Vessel)* **13**, 378–381 (1982).
44. Sakuragawa, N., Shimizu, K., Kondo, K., Kondo, S. and Niwa, M., Studies of the effect of PEG-modified urokinase on coagulation-fibrinolysis using beagles. *Thromb. Res.* **41**, 627–635 (1986).
45. Takahashi, K., Ajima, A., Yoshimoto, T. and Inada, Y., Polyethylene glycol-modified catalase exhibits unexpectedly high activity in benzene. *Biochem. Biophys. Res. Commun.* **125**, 761–766 (1984).
46. Matsushima, A., Okada, M. and Inada, Y., Chymotrypsin modified with polyethylene glycol catalyzes peptide synthesis reaction in benzene. *FEBS Lett.* **178**, 275–277 (1984).
47. Lee, H., Takahashi, K., Kodera, Y., Ohwada, K., Tsuzuki, T., Matsushima, A. and Inada, Y., Polyethylene glycol-modified papain catalyzes peptide bond formation in benzene. *Biotechnol. Lett.* **5**, 403–407 (1988).
48. Inada, Y., Nishimura, H., Takahashi, K., Yoshimoto, T., Saha, A. R. and Saito, Y., Ester synthesis catalyzed by polyethylene glycol-modified lipase in benzene. *Biochem. Biophys. Res. Commun.* **122**, 845–850 (1984).
49. Yoshimoto, T., Takahashi, K., Nishimura, H., Ajima, A., Tamaura, Y. and Inada, Y., Modified lipase having high stability and various activities in benzene, and its re-use by recovering from benzene solution. *Biotechnol. Lett.* **6**, 337–340 (1984).

50. Yoshimoto, T., Nakata, M., Yamaguchi, S., Funada, T., Saito, Y. and Inada, Y., Synthesis of eicosapentaenoyl phosphatidylcholines by polyethylene glycol-modified lipase in benzene. *Biotechnol. Lett.* **11**, 771-776 (1986).
51. Yoshimoto, T., Mihama, T., Takahashi, K., Saito, Y., Tamaura, Y. and Inada, Y., Chemical modification of enzymes with activated magnetic modifier. *Biochem. Biophys. Res. Commun.* **145**, 908-914 (1987).
52. Yoshimoto, T., Ritani, A., Ohwada, K., Takahashi, K., Kodera, Y., Matsushima, A., Saito, Y. and Inada, Y., Polyethylene glycol derivative-modified cholesterol oxidase soluble and active in benzene. *Biochem. Biophys. Res. Commun.* **148**, 876-882 (1987).
53. Yoshimoto, T., Ohwada, K., Takahashi, K., Matsushima, A., Saito, Y. and Inada, Y., Magnetic urokinase: targetting of urokinase to fibrin clot. *Biochem. Biophys. Res. Commun.* **152**, 739-743 (1988).
54. Takahashi, K., Nishimura, H., Yoshimoto, T., Okada, M., Ajima, A., Matsushima, A., Tamaura, Y., Saito, Y. and Inada, Y., Polyethylene glycol-modified enzymes trap water on their surface and exert enzymic activity in organic solvents. *Biotechnol. Lett.* **6**, 765-770 (1984).
55. Takahashi, K., Yoshimoto, T., Ajima, A., Tamaura, Y. and Inada, Y., Modified lipase catalyzes ester synthesis in benzene, substrate specificity. *Enzyme* **32**, 235-240 (1984).
56. Takahashi, K., Yoshimoto, T., Tamaura, Y., Saito, Y. and Inada, Y., Ester synthesis at extraordinarily low temperature of $-3°C$ by modified lipase in benzene. *Biochem. Int.* **10**, 627-631 (1985).
57. Takahashi, K., Ajima, A., Yoshimoto, T., Okada, M., Matsushima, A., Tamaura, Y. and Inada, Y., Chemical reactions by polyethylene glycol-modified enzymes in chlorinated hydrocarbons. *J. Org. Chem.* **50**, 3414-3415 (1985).
58. Takahashi, K., Kodera, Y., Yoshimoto, T., Ajima, A., Matsushima, A. and Inada, Y., Ester-exchange catalyzed by lipase modified with polyethylene glycol. *Biochem. Biophys. Res. Commun.* **131**, 532-536 (1985).
59. Takahashi, K., Tamaura, Y., Kodera, Y., Mihama, T., Saito, Y. and Inada, Y., Magnetic lipase stable and active in organic solvents. *Biochem. Biophys. Res. Commun.* **142**, 291-296 (1987).
60. Ajima, A., Yoshimoto, T., Takahashi, K., Tamaura, Y., Saito, Y. and Inada, Y., Polymerization of 10-hydroxydecanoic acid in benzene with polyethylene glycol-modified lipase. *Biotechnol. Lett.* **7**, 303-306 (1985).
61. Ajima, A., Takahashi, K., Matsushima, A., Saito, Y. and Inada, Y., Retinyl esters synthesis catalyzed by polyethylene glycol-modified lipase in benzene. *Biotechnol. Lett.* **8**, 547-552 (1986).
62. Matsushima, A., Okada, M., Takahashi, K., Yoshimoto, T. and Inada, Y., Indoxyl acetate hydrolysis with polyethylene glycol-modified lipase in benzene solution. *Biochem. Int.* **11**, 551-555 (1985).
63. Matsushima, A., Kodera, Y., Takahashi, K., Saito, Y. and Inada, Y., Ester-exchange reaction between triglycerides with polyethylene glycol-modified lipase. *Biotechnol. Lett.* **8**, 73-78 (1986).
64. Nishio, T., Takahashi, K., Yoshimoto, T., Kodera, Y., Saito, Y. and Inada, Y., Terpene alcohol ester synthesis by polyethylene glycol-modified lipase in benzene. *Biotechnol. Lett.* **9**, 187-190 (1987).
65. Nishio, T., Takahashi, K., Tsuzuki, T., Yoshimoto, T., Kodera, Y., Matsushima, A., Saito, Y. and Inada, Y., Ester synthesis in benzene by polyethylene glycol-modified lipase from *Pseudomonas fragi* 22.39B. *J. Biotechnol.* **8**, 39-44 (1988).
66. Kodera, Y., Takahashi, K., Nishimura, H., Matsushima, A., Saito, Y. and Inada, Y., Ester synthesis from α-substituted carboxylic acid catalyzed by polyethylene glycol-modified lipase from *Candida cylindracea*. *Biotechnol. Lett.* **8**, 881-884 (1986).
67. Tamaura, Y., Takahashi, K., Kodera, Y., Saito, Y. and Inada, Y., Chemical modification of lipase with ferromagnetic modifier: A ferromagnetic-modified lipase. *Biotechnol. Lett.* **8**, 877-880 (1986).
68. Mihama, T., Yoshimoto, T., Ohwada, K., Takahashi, K., Akimoto, S., Saito, Y. and Inada, Y., Magnetic lipase adsorbed to a magnetic fluid. *J. Biotechnol.* **7**, 141-146 (1988).
69. Takahashi, K., Nishimura, H., Yoshimoto, T., Saito, Y. and Inada, Y., A chemical modification to make horseradish peroxidase soluble and active in benzene. *Biochem. Biophys. Res. Commun.* **121**, 261-265 (1984).
70. Yoshimoto, T., Chao, S. G., Saito, Y., Imamura, I., Wada, H. and Inada Y., Chemical modification of tryptophanase from *E. coli* with polyethylene glycol to reduce its immunoreactivity towards anti-tryptophanase antibodies. *Enzyme* **36**, 261-265 (1987).

Multiphase Biomedical Materials, pp. 167–190 (1989)
T. Tsuruta and A. Nakajima (Eds)
1989 VSP.

Chapter 11

Improved drug delivery directed to specific tissue using polysaccharide-coated liposomes

JUNZO SUNAMOTO[1] and TOSHINORI SATO[2]

[1]*Laboratory of Materials Science of Polymers and Artificial Cell Technology, Department of Polymer Chemistry, Faculty of Engineering, Kyoto University, Sakyo-ku, Yoshida Hommachi, Kyoto 606, Japan*
[2]*Department of Industrial Chemistry, Faculty of Engineering, Nagasaki University, Nagasaki 852, Japan*

Summary—In order to prepare a more cell specific and more stable drug carrier for water soluble drugs, an improved methodology has been developed. This methodology involves coating the outermost surface of liposomes with a cholesterol derivative of naturally occurring polysaccharides, such as mannan or amylopectin. Water soluble drugs such as antibiotics or anti-tumor drugs were encapsulated in the water phase of the liposomes. As expected, the polysaccharide-coated liposomes showed excellent mechanical and biochemical stability both *in vitro* and even *in vivo* after administration to the body. In addition, due to the sort of chemical structure of the terminal sugar moiety of the polysaccharides, cell specificity was developed. Using the polysaccharide-coated liposomes, treatments of infectious diseases in animal were tried and great successes were attained.

Furthermore, by chemically conjugating a fragment of monoclonal antibody to the polysaccharide and coating the surface of liposome containing the anti-tumor drug by the modified polysaccharide, we could prepare a new type of immunoliposomes. This newly developed immunoliposomes containing adriamycine also showed great activity in the treatment of tumor-bearing mice.

1. INTRODUCTION

Effective delivery of drugs to target specific cells or tissues brings about a diminish of toxic side-effects and an increase in the pharmacological activity of drugs, and subsequently makes a decrease in the dosage of drugs possible. Targeting of drugs to specific cells or tissues, as well as sustained release, is the most important problem in drug delivery systems (DDS).

More precisely, 'targeting' must be classified into two different categories on the basis of the difference in mechanism *in vivo* after administration. One is 'passive targeting' [1, 2] with a bulk recognition mechanism, in which the targeting (recognition of a specific cell or a tissue) is attained by altering the bulk structural characteristics of the carrier, e.g. the hydrophobicity (or lipophilicity), the charge density, the fluidity (or softness) and/or the size of carrier. Therefore, this mechanism may be more important when the drug carrier and the target cell are a long distance apart from each other. Another mechanism is 'active targeting' which is attained by molecular recognition mechanisms. A recognition site, such as an antibody, an antigen or a saccharide determinant, is covalently or non-covalently assembled on the surface of drug carriers and behaves as an effective sensory device in a missile drug carrier. In this mechanism, the targeting is attained by the recognition at the molecular level through the direct and specific interaction between the recognition site on the surface of the drug carrier and the receptor in membranes of the target cell. Therefore, this targeting mechanism becomes more effective when the carrier and the target cell are closer to each other. This means that the *active targeting* is more eminent and promising than the *passive targeting*.

The liposome has been widely accepted to be a possible drug carrier, especially for water soluble drugs [3]. For water insoluble (lipophilic) drugs we can employ other better carriers such as nanoparticles, various emulsions, polymer capsules, etc. From its less toxicity, low immunogenecity and biodegradability, the liposome certainly seems to be one of the good candidates for a drug carrier. However, it still has two serious disadvantages; namely, the structural instability both *in vitro* and *in vivo*, and reduced targetability or cell specificity. Therefore, we have to overcome these disadvantages of the liposome before practical use.

Since 1982, we have developed an improved methodology for making liposomes more stable and cell specific [4–7]. This methodology involves coating the surface of liposomes by naturally occurring polysaccharides bearing a cholesterol moiety with a suitable spacer. Such a polysaccharide-coated liposome was found to show an increase in stability against external stimuli such as pH, ionic strength, osmotic pressure, temperature and/or polarity of the medium *in vitro*, and the attack by lipases, lipooxidases, and serum proteins *in vivo* [3–5]. In addition, these polysaccharide-coated liposomes, especially amylopectin- or mannan-coated ones, showed a unique specificity to phagocytes such as alveolar macrophages, monocytes and neutrophils [6, 7]. This cell specificity of the mannan-coated liposome was interpreted in terms of the specific interaction between the mannose moiety assembled on the liposomal surface and the mannose-recognizable receptor in the cell membrane [8].

In this chapter, we would like to overview several results recently obtained by ourselves by employing the polysaccharide-coated liposomes.

2. PREPARATION OF POLYSACCHARIDE-COATED LIPOSOMES

Polysaccharides, when partly derivatized by simple palmitoyl moieties, were found to develop [3, 4] stable vesicles. Subsequently we have found that cholesterol derivatized polysaccharides, bearing a suitable spacer, gave a much better coating efficiency of liposomes [9]. These days, we employ only cholesterol substituted polysaccharides (Fig. 1). The preparation of these polysaccharide derivatives bearing cholesterol groups has been previously described [9, 10]. The degree of substitution of the cholesteryl moiety, per hundred monosaccharide units, was determined by ^{1}H-NMR. Amylopectin

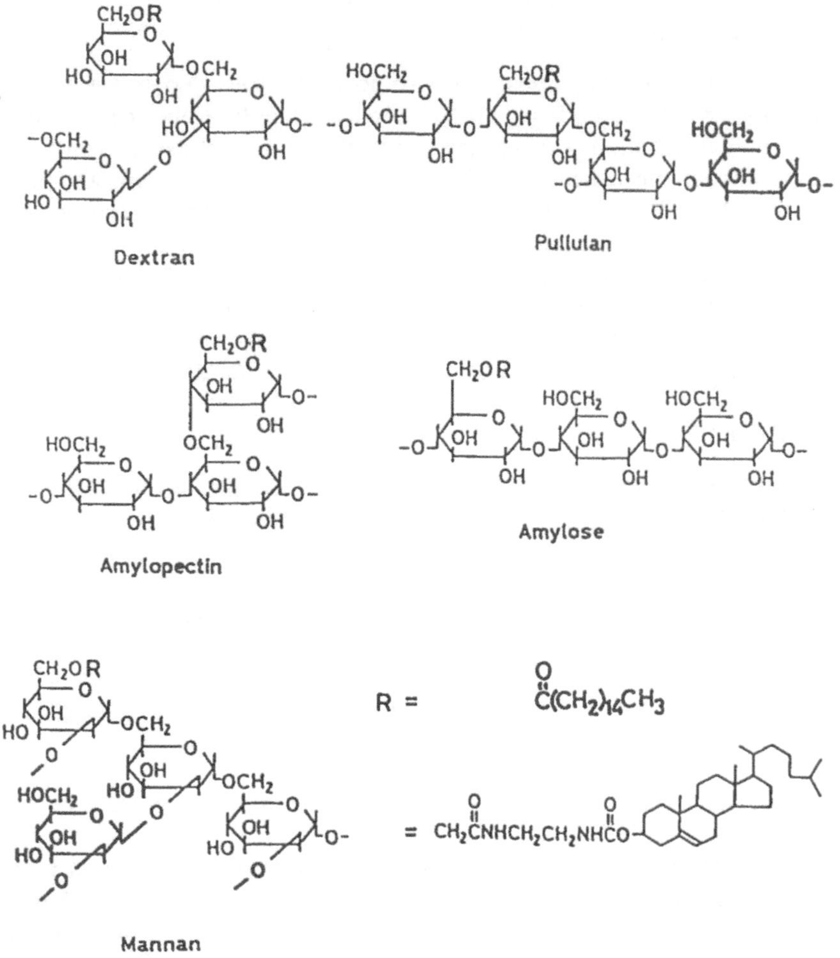

Figure 1. Polysaccharide derivatives used for coating of the liposomal surface.

(molecular weight 112 000) was substituted by 1.0 cholesteryl moiety per 100 glucose units and coded as CHA-112-1.0. Mannan (molecular weight 200 000) was substituted by 1.7 cholesteryl moieties per 100 mannose units and was coded as CHM-200-1.7.

Polysaccharide-coated liposomes were prepared essentially by the same procedure as that adopted for conventional liposomes without any polysaccharide [9, 10]. For example, 1.0 ml of saline containing 7.5 mg of polysaccharide derivative was added to 4.0 ml of a liposome suspension which contained 37.5 mg of egg lecithin. After stirring for 30 min below 20.0°C, the resulting mixture was gel-filtered by passing through a Sepharose 4B column. LUVs (large unilamellar vesicles) with or without polysaccharide derivatives were prepared by the reverse phase evaporation technique [11].

3. EFFICIENCY OF COATING LIPOSOMES WITH POLYSACCHARIDE DERIVATIVES

The efficiency of coating liposomes with polysaccharide derivatives was ascertained by several methods [4, 6]: (i) isolation of polysaccharide-coated liposomes by gel-filtration, (ii) reduced permeability of carboxyfluorescein (CF) encapsulated in the interior of the liposomes with or without serum and plasma, (iii) increased resistance to phospholipase and lipoxygenase, (iv) enzymatic digestion of polysaccharides adsorbed on the surface of liposomes and (v) specific aggregation of polysaccharide-coated liposomes with concanavalin A. All of these tests showed that the liposomes were effectively coated with polysaccharide derivatives and that these artificial cells, covered by an artificial polysaccharide cell wall, could be used as an improved drug carrier.

4. TARGETING

Liposomes are usually cleared rapidly by macrophages in the reticuloendothelial system (RES), which sometimes causes difficulty in the utilization of liposomes as drug carriers. However, in the delivery of drugs such as immunomodulators, interferon inducers or antibiotics to macrophages, this can be an advantage. In these cases, the macrophages could serve as processing cells that would allow the appropriate presentation of certain antigens to the immunosystem. In fact, Fidler et al. [12, 13], Sone et al. [14] and Phillips et al. [15] have shown that liposome encapsulated muramyl dipeptide (MDP) exhibits a much stronger tumoricidal activity in comparison with free MDP.

4.1. Effective internalization of polysaccharide-coated liposomes by phagocytes

Internalization of liposomes by phagocytic cells was investigated quantitatively by both flow cytofluorometry and radioisotope techniques. A 1.0 ml cell suspension containing 1.5×10^6 cells was incubated with 5.0 or 50.0 μl of a liposome suspension (30.0 mg egg PC per 4.0 ml) for 30 min at 37.0°C in RPMI-1640 containing 10% heat-inactivated FCS. For the flow cytofluorometry, CF-loaded liposomes were prepared and mixed with the cell suspensions. After incubation the fluorescence was measured on a FCM-1S spectrometer. The results obtained are shown in Fig. 2. For the radioactive investigation, [^{14}C]dipalmitoylphosphatidylcholine as a probe was incorporated in the lipid membrane. After incubating the phagocytes with these 'hot' liposomes, the resulting cell suspension was centrifuged first for 10 min at 400 g and then at 10 000 g for 3 min in a velocity-gradient silicon oil. The radioactivities of both the sediment (cells) and supernatant were determined on a Aloka III scintillation counter. The number of liposomes internalized by each phagocyte was estimated by assuming that the average diameter of the multilamellar vesicles (MLV) was 109 nm. The results are illustrated in Fig. 3.

It is very clear from the data listed in Figs. 2 and 3 that the alveolar macrophages from guinea pigs as well as monocytes and neutrophils from human blood exhibit a remarkable enhancement in the internalization when the

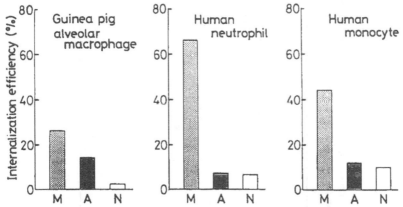

Figure 2. Internalization efficiency of liposomes into three kinds of phagocytic cells monitored by flow cytofluorometry using CF-loaded liposomes. Liposomes (5.2×10^{10} LUVs/ml) were incubated with phagocytes (2.0×10^{6} cells/ml) at 37.0°C for 60 min. M, mannan-coated LUV; A, amylopectin-coated LUV; N, non-coated LUV.

liposomes are coated with amylopectin or mannan compared to conventional liposomes without these polysaccharide coatings.

4.2. *Tissue distribution of mannan-coated liposomes*

Table 1 shows the tissue distribution of radioactivity after intravenous injection of the cholesteryl mannan-coated and conventional liposomes, as monitored by the radioisotope technique. Substantial amounts of both isotopes were found in the lung and liver at 30 min after injection of the cholesteryl mannan-coated liposomes. As expected from the high clearance rate of cholesteryl mannan-liposomes, the amounts of [^{3}H]inulin and [^{14}C]coenzyme Q_{10} in the lung plus

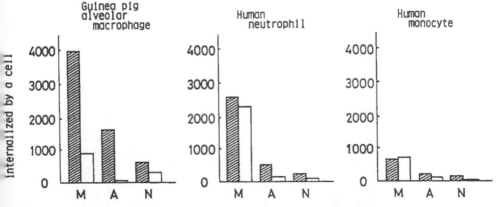

Figure 3. Internalization efficiency of liposomes into three kinds of phagocytic cells monitored by RI method using [^{14}C]DPPC labeled liposomes. Liposomes (▨, 3.9×10^{11} LUVs/ml and ☐, 3.9×10^{10} LUVs/ml) were incubated with phagocytes (1.5×10^{6} cells/ml) at 37.0°C for 60 min. M, mannan-coated LUV; A, amylopectin-coated LUV: N, non-coated LUV.

Table 1. Tissue distribution of [³H]inulin and [¹⁴C]coenzyme Q_{10} after intravenous injection of conventional liposomes or cholesteryl mannan-coated liposomes. Guinea pigs were injected intravenously with 0.25–0.3 ml of a liposomal suspension containing 3.1–3.7 μmol phopholipid. Data from three animals are mean ± S.E.

Tissue	Time after injection (h)	Percentage of the dose in each tissue					
		Conventional liposome			Cholesteryl mannan-coated liposome		
		[³H]inulin	[¹⁴C]coenzyme Q_{10}	³H/¹⁴C	[³H]inulin	[¹⁴C]coenzyme Q_{10}	³H/¹⁴C
heart	0.5	0.37 ± 0.12	0.43 ± 0.09	0.9	0.20 ± 0.00	0.27 ± 0.03	0.7
	24	0.09 ± 0.03	0.38 ± 0.12	0.2	0.03 ± 0.003	0.20 ± 0.00ᵃ	0.2
lung	0.5	3.1 ± 0.9	30.9 ± 6.8ᵇ	0.1	34.9 ± 3.0	67.1 ± 4.3ᵃ	0.5
	24	0.22 ± 0.06	5.6 ± 0.6ᵃ	0.04	0.87 ± 0.17	12.5 ± 2.1ᵃ	0.07
spleen	0.5	7.0 ± 1.1	10.8 ± 3.8	0.6	1.4 ± 0.2	2.8 ± 0.3ᵇ	0.5
	24	20.8 ± 3.5	19.5 ± 3.7	1.1	3.2 ± 0.4	7.9 ± 0.9ᵇ	0.4
liver	0.5	4.5 ± 0.2	22.2 ± 1.6ᵃ	0.2	10.3 ± 0.7	24.3 ± 1.5ᵃ	0.4
	24	16.8 ± 2.1	53.9 ± 0.8ᵃ	0.3	21.4 ± 0.8	54.3 ± 2.2ᵃ	0.4
kidney	0.5	2.3 ± 0.3	0.63 ± 0.09ᵃ	3.6	6.8 ± 3.3	0.4 ± 0.06	17.0
	24	0.72 ± 0.14	0.31 ± 0.06	2.3	1.4 ± 0.2	0.23 ± 0.03ᵃ	6.1
blood	0.5	20.1 ± 2.5	13.8 ± 1.9	1.5	9.2 ± 0.8	2.8 ± 0.7ᵃ	3.3
	24	0.49 ± 0.16	1.2 ± 0.1ᵇ	0.4	0.34 ± 0.04	1.1 ± 0.2ᵇ	0.3

ᵃ Significantly different from [³H]inulin (P < 0.01).
ᵇ Significantly different from [³H]inulin (P < 0.05).

liver, after the injection, were also high, being 45.2 and 91.4% of the dose, respectively. The lung-uptakes of the polymer-coated liposomes for [^{3}H]inulin and [^{14}C]coenzyme Q_{10} were approximately 11- and 2-times higher than those of the control liposomes, respectively. Even at 24 h after injection of the polymer-coated liposomes, these values in the lung were still higher than those for the conventional liposomes: 4-times for ^{3}H and 2-times for ^{14}C, respectively. However, the value in the liver was almost the same for both the conventional and the polysaccharide-coated liposomes. The uptake of the cholesteryl mannan-coated liposomes in the spleen at 30 min after injection was significantly decreased compared with those of the conventional liposomes: approximately 1/5 for [^{3}H]inulin and 1/4 for [^{14}C]coenzyme Q_{10}, respectively.

For the cholesteryl mannan-coated liposomes, the ^{3}H/^{14}C ratios in the lung and liver at 30 min after injection were less than unity, but were approximately 5- and 2.5-times higher than those for the conventional liposomes. Conversely, the ratios in the kidney and blood were higher than unity, but remained higher than those of the conventional liposomes, being approximately 5- and 2-times, respectively. At 24 h after injection, the ^{3}H/^{14}C ratios in the lung and blood became less than unity. However, the values remained higher than those of the control. Meanwhile, the ratio in the liver was almost comparable with that of the control. The ^{3}H/^{14}C ratios in the lungs and liver after the injection were higher in the cholesteryl mannan-coated liposomes than in the conventional liposomes. This higher ratios implies that the cholesteryl mannan-coated liposome is more stable even *in vivo* than the conventional liposome. The ratio in the kidney was also more than unity, and the value higher than that of the control was kept even after 24 h. This result might be due to leakage of [^{3}H]inulin into the blood upon its liberation from the cholesteryl mannan-coated liposomes, followed by its accumulation in the kidney prior to excretion into the urine. In the spleen, on the other hand, the ^{3}H/^{14}C ratio for the cholesteryl mannan-coated liposomes was lower than that of the control experiment. This result suggests that the cholesteryl mannan-coated liposomes are more easily destructed in the spleen than the conventional liposome. Specifically, in the case of the control, 20.8% of the administrated [^{3}H]inulin still remained in the spleen even 24 h after the injection. For the cholesteryl mannan-coated liposomes, however, the value was 3.2%. Comparison of the cumulative amounts of urinary [^{3}H]inulin at 24 h after the injection showed that the value for the conventional liposomes was less than that for the cholesteryl mannan-coated liposomes.

5. APPLICATION OF POLYSACCHARIDE-COATED LIPOSOMES IN CHEMOTHERAPY AND IMMUNOTHERAPY

5.1. *Application of the polysaccharide-coated liposomes to chemotherapy*
The best treatment of bacterial infectious diseases involves the use of antibiotics. It is logical, therefore, to consider liposomes as carriers of antibiotics to phagocytic cells, which in certain infections harbor microorganisms which can survive intracellularly. Antibiotics can act also against such intracellular infections if they can penetrate the phagocyte.

In 1976, a new bacterial infectious disease, Legionnaires' disease, was found
in Philadelphia, USA. There are several problems in the treatment of Legionnaires' disease with antibiotics: the bacterium, *Legionella pneumophila*, grows
even in macrophages and produces β-lactamase. OPA-112-(4.9)-coated LUV in
which an aminoglycoside antibiotics (sisomycin; Sherring, USA) is encapsulated, has been employed for the treatment of experimental Legionnaires'
disease in guinea pigs [6, 7].

5.1.1. Multiplication of L. pneumophila *in human monocytes.* Prior to
starting the treatment, multiplication of *L. pneumophila* in human monocytes
was investigated using sisomycin-encapsulated liposomes. The result is shown
in Fig. 4. With the exception of the control (A) and the experiment (B, free
antibiotics were administered), 36% of the antibiotics administered were encapsulated in the polysaccharide-coated LUV. This means that 64% of antibiotics
were free of liposome encapsulation and were able to kill bacteria present at the

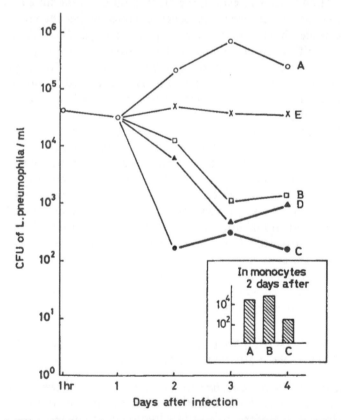

Figure 4. Effect of sisomycin encapsulated in OPA-coated LUV on *Legionella pneumophila*
multiplication in human monocytes at 37.0°C: A (O), control experiment without antibiotics; B
($\square$), free antibiotics, 50 μg/ml, C ($\bullet$), 50 μg/ml, in the presence of liposomes; D ($\blacktriangle$), 5 μg/ml, in the
presence of liposomes; and E ($\times$), 0.5 μg/ml, in the presence of liposomes. When the OPA-coated
LUVs were employed, 36% of the total dose were encapsulated in liposomes and the rest was free.

exterior of cells. The most noteworthy finding is the CFU in monocytes (see insertion of Fig. 4). Even if 50 μg/ml of sisomycin were administered, free sisomycin did not kill the bacteria in the cells, and no significant difference in CFU in monocytes was observed in comparison with the control. On the other hand, the liposome-encapsulated antibiotics showed a drastic decrease in CFU in the cells.

In addition, multiplication of *Staphylococcus* and *Branhamella catarrhalis* has been investigated by employing OPA-112-(4.9)-coated liposomes which contain gentamicin sulfate, piperacillin or ceftazidime. In all the cases studied, when the antibiotics were encapsulated by the polysaccharide-coated liposomes, bactericidal activity of these antibiotics was largely enhanced.

5.1.2. Treatment of experimental Legionnaires' disease in guinea pigs [6]

According to the method established by Pennington *et al.* in 1978, intratracheal inoculation with tracheotomy was carried out in guinea pigs (body wt 280–320 g). Treatment was started 24 h after inoculation and 4 mg/kg/day of the antibiotics was administered by intravenous or intramuscular injection twice a day for 7 days. In this treatment, 36% of the antibiotic was encapsulated in OPA-112-(4.9)-coated LUV and the rest was free. When the disease was treated with free antibiotics without liposomes, all the guinea pigs died within 6 days, while 100% survival rate was attained when the animals were treated with a mixture of free and liposome-encapsulated antibiotics (Fig. 5). To our knowledge, this was the first successful demonstration of the treatment of bacterial infectious disease in the receptor-mediated targeting with antibiotics-bearing liposomes [6].

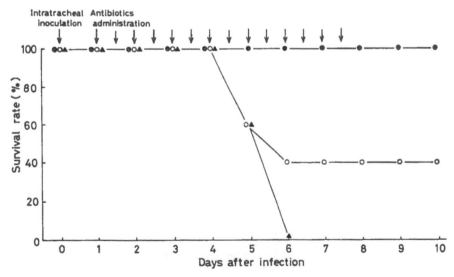

Figure 5. Effect of sisomycin encapsulated in OPA-coated LUV in the treatment of experimental Legionnaires' disease in guinea pigs: ▲, antibiotic without liposomes was administered intramuscularly; O, antibiotics (of which 36% was encapsulated in OPA-coated LUV) was administered intramuscularly; ●, antibiotic (of which 36% was encapsulated in liposomes) was administered intravenously. Five guinea pigs were treated in each experiment.

5.1.3. Treatment of experimental pulmonary candidiasis in mice [16]. Pulmonary candidiasis was treated with the polysaccharide-coated liposomes containing Amph.B (amphotericin B) or free Amph.B. Amph.B (0.2–2.0 mg/ml) was encapsulated in multilamellar vesicles (egg phosphatidylcholine, 15.0–30.0 mg/ml), which were coated with amylopectin (1.5–30.0 mg/ml). Groups of 10 normal mice were injected with free Amph.B (0.8, 1.2 and 2.0 mg/kg) and amylopectin-coated liposome containing Amph.B (2.5, 5.0 and 10.0 mg/kg), and the survival rate was determined.

BALB/c mice were inoculated intratracheally with *Candida albicans* (6.4 × 10^6 CFU) and 2 h later amylopectin-coated liposomes containing Amph.B or

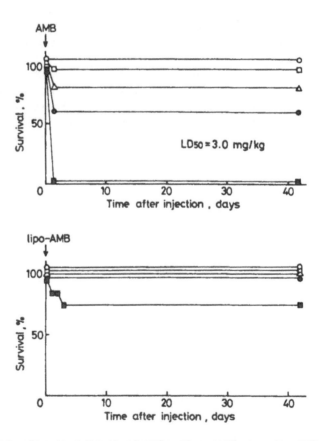

Figure 6. Toxicity of free Amph.B (top) and amylopectin-coated liposome ([egg PC] = 60 mg/kg) containing Amph.B (bottom) to normal mice (*n* = 20). Top; (○), Amph.B 0.8 mg/kg; (□), Amph.B 1.2 mg/kg; (△), Amph.B 2.0 mg/kg; (●), Amph.B 3.0 mg/kg; (■), Amph.B 6.0 mg/kg. Bottom; (○), empty liposome; (□), lipo-Amph.B 1.0 mg/kg; (△), lipo-Amph.B 2.5 mg/kg; (●), lipo-Amph.B 5.0 mg/kg; (■), lipo-Amph.B 10.0 mg/kg.

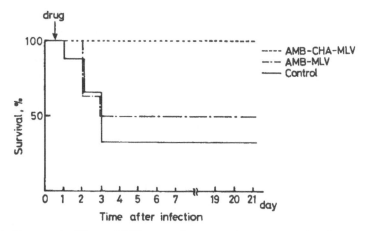

Figure 7. Therapeutic efficacy of CHAp-coated or conventional multilamellar liposomes containing Amph.B (5 mg/kg) in mice infected with *C. albicans*.

free Amph.B (5 mg/kg) were injected once into the tail vein. The toxicities of free Amph.B and amylopectin-coated liposome-contained Amph.B were examined in mice (Fig. 6). The LD_{50} of Amph.B is 1.2 mg/kg, but when amylopectin-coated liposome bearing Amph.B was administered all mice survived, even at a dose of 5 mg/kg. This indicates a remarkable reduction in toxicity of Amph.B by employing the polysaccharide-coated liposome delivery system.

The survival rates of mice with *pulmonary candidiasis* treated with amylopectin-coated liposome and conventional liposome containing Amph.B are shown in Fig. 7. The amylopectin-coated liposome showed remarkable efficacy (100%), which was much better than the conventional liposome (50%) even at the same dosage.

5.2. Application of polysaccharide-coated liposomes to immunotherapy
One difficulty in the utilization of liposomes as drug carriers is that liposomes are cleared rapidly by macrophages in the RES. However, delivery of drugs such as immunomodulators and interferon inducer to macrophages is not a disadvantage. Activated macrophages may actively participate in the killing of bacteria or tumor cells. Macrophages which are activated by a variety of immunopotentiators such as interferon, lymphokines, muramyldipeptide [12–15] or polyanionic polymers [17] show cytotoxicity. However, the *in vivo* activation of macrophages by injection of these agents has not been very successful, presumably due to both the lack of targetability of the free agent to macrophages and the fast clearance of the agent from the blood stream.

In order to modulate the *in vivo* adjuvant activity of an immunopotentiator, encapsulation of immunopotentiators into liposomes has been attempted, and several successful results were obtained [18–20].

5.2.1. Muramyldipeptide (MDP). N-Acetylmuramyl-L-alanyl-D-iso-gluta-min (MDP) is considered to be a minimum structure possessing immunoadju-vant activity which is responsible for mycrobacteria or Freund's complete adjuvant [21]. Since MDP is isolated from peptideglycan, a large number of analogues or derivatives have been provided and evaluated for biological activity [22].

By employing liposomes as a vehicle to deliver MDP into macrophages, the tumoricidal activity was considerably enhanced [12, 14, 15]. However, success-ful results for delivery of liposomal MDP to alveolar macrophages were very few *in vivo* [12]. We employed mannan-coated liposomes, containing a MDP derivative, N^α-(N-acethylmuramyl-L-alanyl-D-iso-glutaminyl)-N^ϵ-stearyl-L-lysine, MDP-L18 (Daiichi Seiyaku, Japan), to target and activate alveolar macrophages of C57BL/6 mice *in vitro* and *in vivo*.

Cytotoxic activity of the effector cells, which were alveolar macrophages (AMϕ, 2×10^5) activated by free MDP or liposomal MDP, was evaluated by ^{51}Cr release from the target cell, Lewis lung carcinoma (3LL, 1×10^4). Cyto-lysis was determined by following equation:

% Cytolysis

$$= \frac{\text{cpm(activated AM}\phi + \text{target cell)} - \text{cpm(normal AM}\phi + \text{target cell)}}{\text{cpm(maximum release)} - \text{cpm(normal AM}\phi + \text{target cell)}} \times 100$$

When AMϕ was activated *in vitro*, the enhancement of the tumoricidal activity of AMϕ activated by liposomal MDP was not significant compared to the free MDP. However, when AMϕ was activated *in vivo* for 18 h by intravenous injection of liposomal MDP to mice, a drastic increase in the tumoricidal effect of AMϕ was observed (Fig. 8).

5.2.2. Polyanionic polymers. Several polyanionic polymers such as pyran and poly(acrylic acid-alt-maleic acid) have been found to exhibit a wide range of biological activities *in vivo* [17]. In particular, they show an ability to enhance macrophage function including inhibition of tumor growth. Recently, Otten-brite and his collaborators have reported that the hydrolyzates of alternate copolymers of maleic anhydride with 2-cyclohexyl-1,3-dioxap-5-ene, itaconic acid or styrene show relatively high cytotoxicity against 3LL [23]. Among various polyanionic polymers, poly(maleic acid-alt-2-cyclohexyl-1,3-dioxap-5-ene) (MA-CDA) was one of the better candidates, and was chosen for encap-sulating into the liposomes because it was very potent in macrophage activation and gave no damage to liposomal membranes [18].

First, the effect of macrophage activation was evaluated by monitoring superoxide production from peritoneal exudate cells (PEC) of C57BL/6 mice collected after intraperitoneal injection of MA-CDA (as free or encapsulated in liposomes). Superoxide anion production was determined by the cytochrome c method.

MA-CDA showed greater activity in the cytotoxic assay than pyran. Production of O_2^- from PEC (2×10^5 cells/ml) upon activation by the

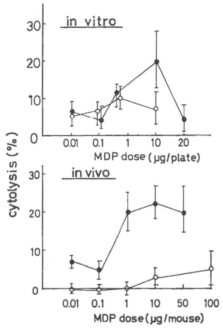

Figure 8. Cytotoxic activity of alveolar macrophages activated by free MDP-L18 (O) or MDP-L18 encapsulated in CHM-coated LUV (●) against cultured 3LL at $E/T = 20$.

immunopotentiator (500 µg/100 µl) was measured 24 h after intraperitoneal (i.p.) injection to mice, while the cytotoxicity of macrophage, which was evaluated by killing of cultured tumor cells, appeared later than 16 h after injection of the drug. Observed O_2^- production induced by MA-CDA was 1.91 ± 0.26 nmol/min/2 × 10^5 cells, and greater than that (1.38 ± 0.12 nmol/min/ 2 × 10^5 cells) by pyran. Results obtained are closely correlated with their cytotoxic activity previously investigated [23].

Superoxide anions liberated from peritoneal exudate macrophages (PEM, 2 × 10^5 cells/ml) activated by MA-CDA (500 µg/100 µl) were followed kinetically (Table 2). In all the runs, the extent of superoxide anion production was always normalized to be 100% as control when only PBS was injected. In the case of free MA-CDA, significant superoxide anion production was observed 5 h after i.p. injection. In the case of MA-CDA encapsulated in either conventional or mannan-coated liposomes, on the other hand, superoxide anion production appeared 30 min after injection. For the liposomal systems, in addition, the maximum increase was observed at 2 h after injection. The increase in superoxide production 2 h after injection of MA-CDA in CHM-LUV (315%) was greater than that of MA-CDA in conventional LUV (250%). Diminution of superoxide anion production induced by MA-CDA encapsulated in the conventional liposome was faster than that for MA-CDA encapsulated in CHM-LUV. Clearly, these data show that MA-CDA can enter more rapidly and effectively into PEM by encapsulation in the CHM-coated liposome

Table 2. Kinetics of superoxide anion production from mouse peritoneal macrophages activated by MA-CDA[a]

Incubation time (h)	Percentage increase in O_2^- production		
	free MA-CDA	MA-CDA in CHM-LUV[b]	MA-CDA in conventional LUV[c]
0.05	101	100	101
0.5	110	152	156
1	100	233	227
2	92	315	250
5	187	182	245
24	201	141	104

[a] Values are calculated by comparing with those of control without any treatment (100%).
[b] CHM-LUV stands for large unilamellar liposomes coated by cholesterol moiety-bearing mannan.
[c] [Egg PC] = 0.75 mg/mouse; [MD-CDA] = 500 μg/mouse; [CHM-200-1.7] = 0.15 mg/mouse.

compared with the cases of free and encapsulation into the conventional liposomes. Also, when only CHM-LUV without MA-CDA was administered, we observed significant superoxide anion liberation from PEM. In this case, however, superoxide anion production was not correlated with the cytotoxicity of macrophages and only shrinkage of macrophages was observed. This was also confirmed by electron microscopic observation of the incubated cells (*vide infra*).

Secondly, tumoricidal properties in mouse alveolar macrophages activated by intravenous injection of MA-CDA was investigated [24]. Growth inhibition for 3LL (at $E/T = 20$) was estimated by [³H]thymidine uptake after culturing of 3LL in the presence of AMϕ for 48 h. The percentage cytostasis was calculated from the following equation:

$$\% \text{ cytostasis} = \frac{\text{cpm(normal AM}\phi + 3\text{LL}) - \text{cpm(activated AM}\phi + 3\text{LL})}{\text{cpm(3LL alone)}} \times 100$$

Cytostatic activity was investigated as a function of incubation time at two different concentrations of MA-CDA (Fig. 9). When 500 μg of MA-CDA encapsulated in CHM-coated liposomes was injected into mice, the maximum tumoricidal activity was observed at 5 days after administration and increased up to 22.5%. Furthermore, when 1000 μg of MA-CDA encapsulated in CHM-coated liposome was injected, the maximum tumoricidal activity was 76% at 3 days after administration, while it was approximately 20% when free MA-CDA was administered.

Finally, activation of alveolar macrophages was morphologically investigated by scanning electron microscopy (SEM). AMϕ was collected by the broncheo alveolar lavage (BAL) method at 3 days after administration of CHM-coated liposomes (vacant or containing MA-CDA), and they were incubated for 18 h with or without target tumor cells. Thereafter, the cell suspensions were submitted to SEM observation. Interestingly, AMϕ treated with vacant liposomes just shrank and showed no tumoricidal activities (Fig. 10a). AMϕ treated

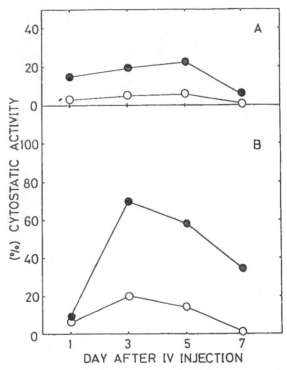

Figure 9. Kinetics of cytostatic activity of mouse alveolar macrophages activated by i.v. injection of MA-CDA against 3LL. Data represents cytostatic activity at an E/T ratio = 20. (●); MA-CDA in CHM-200-1.8 coated LUV, (○); free MA-CDA. Dose of MA-CDA is 500 μg/mouse (A) and 1000 μg/mouse (B).

with CHM-coated liposome containing MA-CDA, on the other hand, significantly exposed pseudopodia and attacked target tumor cells (Fig. 10b).

6. NEWLY DEVELOPED IMMUNOLIPOSOMES

6.1. Strategy

Many approaches have been tried in order to obtain targetability of liposomes in drug delivery systems. These approaches include non-covalent association of cell-specific antibodies with liposomes, covalent attachment of a polyclonal or monoclonal antibody to liposomes, the use of glycoprotein-bearing liposomes or natural and synthetic glycolipid-containing liposomes [7]. The following factors are required for the preparation of an efficient targetable liposome: (i) a sufficient quantity of the sensory device must be bound to the liposomal surface; (ii) the binding between liposomes and the sensory device must be stable; (iii) the targetable property of the sensory device should remain unchanged even after its conjugation to the liposomes; (iv) the integrity of the liposomes should be preserved during the process of conjugation and before their arrival at the target cell or tissue after *in vivo* administration; and finally

(a)

(b)

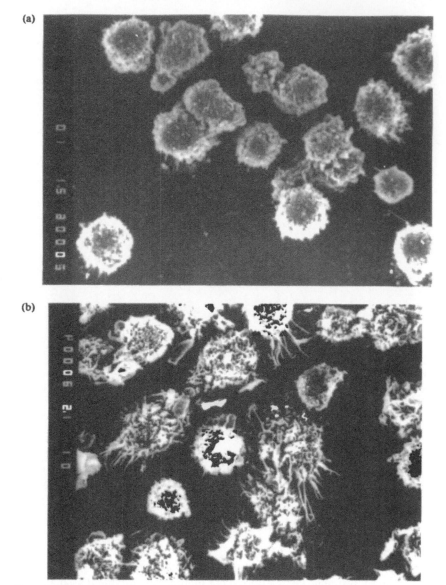

Figure 10. Electron micrograph of alveolar macrophages collected from normal mice treated with vacant CHM-coated LUV (a) or MA-CDA-loaded CHM-coated LUV (b).

(v) the drug-bearing liposomes are required to be effectively internalized into the target cells after binding to them. Although many approaches have been tried, previous techniques to bind a sensory device onto a liposomal surface have not been able to satisfy all of the above requirements. While many basic experiments have been performed, unfortunately only a few has been carried out *in vivo* [25, 26].

On the basis of our successful results in previous studies, we have developed a new and improved technique for safely binding the sensory device, a mono-clonal antibody fragment, onto the liposomal surface [10, 27]. This method involves coating the outer surface of large liposomes with the polysaccharide derivatives and the subsequent covalent binding of the SH bearing antibody fragment to the pullulan derivative on the liposomal surface.

6.2. *Conjugation of the antibody fragment to the pullulan derivative*

For this purpose, pullulan (for example, MW 5×10^4) was selected, because the cholesterol derivative of pullulan can stabilize the liposome with high efficiency, and does not show any significant targetability. Conjugation of an antibody fragment to the pullulan derivative has been carried out basically in accordance with the procedures developed by Hashimoto and his co-workers [28], where the thiol-reacting maleimide group was first introduced into a lipid molecule and then the SH-bearing fragment of an antibody was coupled on the liposomal surface. Different from their method, we first introduced the thiol-reacting maleimide group to the AECM (aminoethylcarbamoylmethyl) group of chol-esteryl pullulan, then coated the outermost surface of egg lecithin large oligo-lamellar vesicles with the polysaccharide so obtained, and finally it coupled with the antibody fragment, which had the SH group, on the liposomal surface. The chemical processes of conjugation of the monoclonal antibody fragment to pullulan and a schematic representation of the immunoliposome formed are illustrated in Figs 11 and 12.

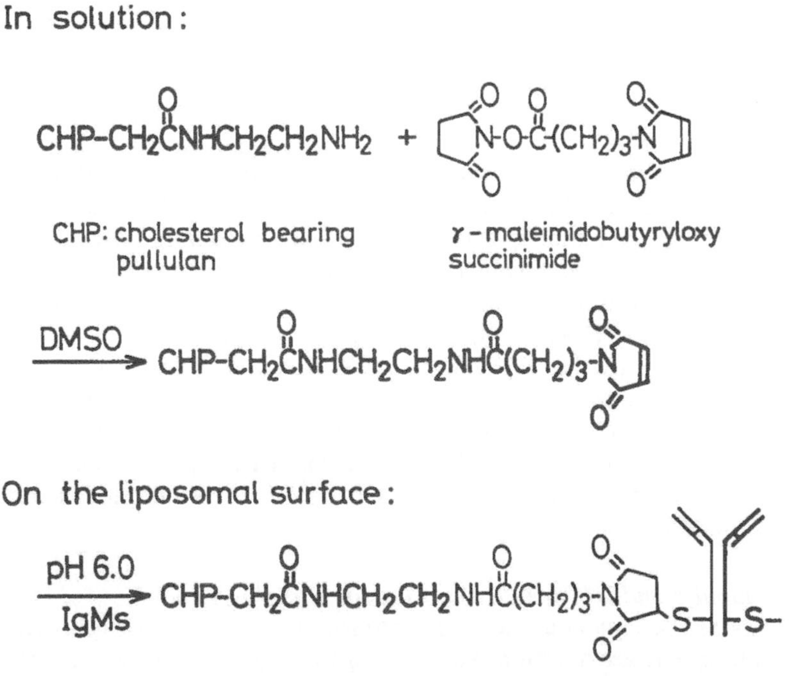

In solution:

CHP: cholesterol bearing pullulan

γ-maleimidobutyryloxy succinimide

On the liposomal surface:

Figure 11. Chemical process for conjugation of the IgMs fragment to cholesterol moiety-bearing pullulan.

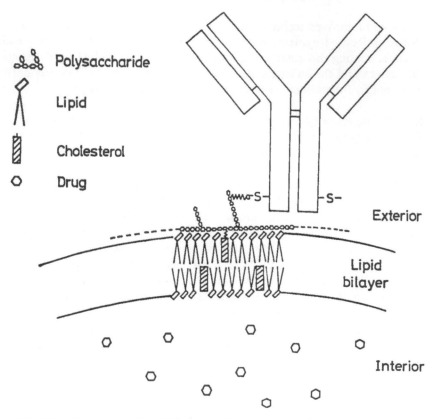

Figure 12. Schematic representation of the immunoliposome prepared.

After conjugation, the immunoliposome was isolated by gelfiltration on a Sepharose 4B column, the separation was monitored by radiochemical analyses of both ^{125}I on the antibody fragment and ^{14}C on the lipid. The gel-filtration showed obvious overlapping of the fraction of the antibody fragment and the liposome. As the first experiment, we have employed the fragment of anti-sialosylated Lewis^x (CSLEX 1), IgMs, which is specific to PC-9 and KATO III and less specific to 3LL [29]. When IgMs was incubated with the pullulan-coated liposomes bearing no coupling device for IgMs, the extent of non-covalent binding of the protein was negligible (less than 2.5%). The average efficiency of conjugation, which was determined from both the amount of recovered protein and the radioactivity of ^{125}I in the liposome fraction, was approximately 50%.

6.3. Stability and cell specificity of our immunoliposome

Stability of liposomes during the procedure of conjugation of IgMs fragment and subsequent gel-filtration was investigated by monitoring the leakage of adriamycin encapsulated in the interior of the liposomes. For example, when adriamycin was encapsulated in the pullulan derivative-coated liposomes and

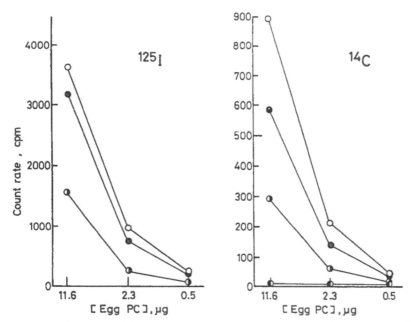

Figure 13. Specific binding of the immunoliposome to various cells (O, PC-9; ●, KATO III; ◑, 3LL) and non-specific binding of the pullulan-coated liposome (◐) as a function of lipid concentration. Liposome was doubly labelled with both [^{14}C]DPPC and [^{125}I]IgMs.

IgMs were conjugated to pullulan on the surface of liposome, 93% of encapsulated adriamycin was retained. On the other hand, when adriamycin-bearing liposomes without any polysaccharide coat were incubated without IgMs under the controlled conditions, the loss of adriamycin was approximately 26%. Further, when the conventional liposome loading adriamycin was co-incubated with the protein, the loss of adriamycin increased up to 63%.

Figure 13 shows relative amount of the immunoliposome bound to the tumor cells, estimated from the count rates of ^{14}C and ^{125}I of the cell-pellets after incubation for 60 min at 37.0°C. The conventional liposome, without the sensory device, did not show any significant binding to any of the cell lines. Conversely, compared with the conventional liposome, the binding of the immunoliposome to specific cells was dramatically increased by factors as much as 447 to PC-9 and 295 to KATO-III, but only by a factor of 148 to the less specific cell 3LL. The result coincides with that for the parent IgM and IgMs in the absence of the liposome.

This specific binding of the immunoliposome to the target cell (PC-9) was further confirmed by fluorescence microscopic observation, employing the hydrophobic fluorescent probe, Tb(acac)$_3$, which was embedded in liposomal membranes. The cell surface of PC-9 was strongly stained by the fluorescent probe. In the case of the anti-CEA Fab' fragment bearing immunoliposome, the liposome is certainly internalized into CEA-producing cells. This was directly confirmed by the fluorescence microscopic technique [27].

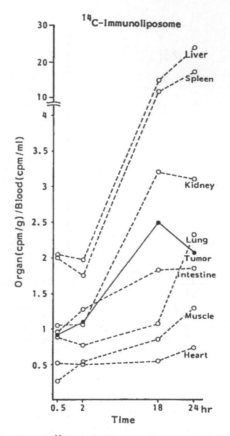

Figure 14. Tissue distribution of [^{14}C]labelled immunoliposome in PC-9-bearing mice.

6.4. Tissue distribution of the immunoliposome [30, 31]

The tissue distribution of the immunoliposomes has been compared with that of GMB-CHP-coated liposomes. When the tumor size of the inoculated athymic mice, which was subcutaneously implanted with human lung cancer PC-9 (1×10^7), became approximately 10–15 mm in diameter, a given amount of ^{14}C-labelled immunoliposomes or GMB-CHP-coated liposomes was intravenously injected, and mice were sacrificed at 30 min, 2 h, 18 h and 24 h after inoculation. The serial ratios of tumor and tissues to blood are given in Fig. 14. The maximum ratio of tumor to blood was observed after 18 h and it was 2.5. The immunoliposome accumulated less in RES at both 30 min and 2 h after inoculation compared with GMB-CHP-coated liposome. Furthermore, the targetability of the immunoliposomes to tumor was compared to that of GMB-CHP-coated liposomes on the basis of the ratio of the radioactivity per 1 g of tumor to that per 1 g of liver. At 30 min and 2 h after administration, the ratios were 0.04 and 0.08, for the GMB-CHP-coated liposomes, while they were 0.37 and 0.57 for the immunoliposomes.

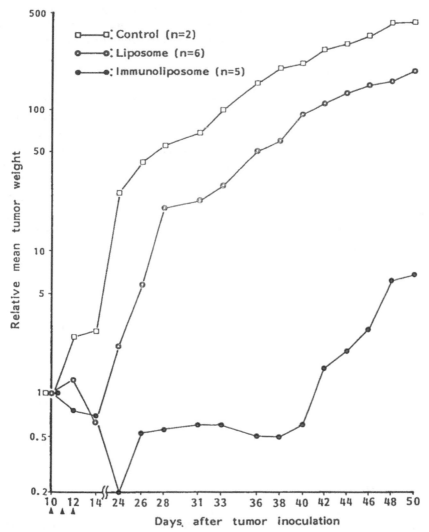

Figure 15. Anti-tumor effect of adriamycin encapsulated immunoliposomes on the growth of implanted PC-9 tumor

6.5. Anti-tumor effect of immunoliposomes bearing ADR in vivo [31]

LUV were employed in order to increase the encapsulation efficiency of water soluble ADR. For example, two groups of five or six mice, which carried pre-existing human lung tumor xenografts of the PC-9 cell line, were treated with both GMB-CHP-coated liposomes and immunoliposomes bearing adriamycin (ADR). The results are shown in Fig. 15, in which changes in relative mean tumor weight are plotted as a function of the time elapsed after the first injection of liposomes. In the case of immunoliposomes, clearly, a statistically significant inhibition of tumor growth ($P < 0.05$) was observed after administration. In the administration of immunoliposomes, there was a gradual reduction in tumor size until 2 weeks, and at this point the tumor was regressed completely. However, the rapid growth started again 28 days after the treatment.

7. CONCLUSION

In order to utilize liposomes as an improved drug carrier, both the assembly of a recognition site on the liposomal surface and increasing the mechanical and chemical stabilities must be attained simultaneously. The most reasonable approach to this problem is to assemble an artificial cell wall on the artificial cell surface. Naturally occurring polysaccharides may be one of the candidates as the artificial cell wall from the view point of their biodegradability, relatively low toxicity and the requirement of the terminal sugar moiety in the biological recognition process. Very recently, we have succeeded in the preparation of specific liposomes which have a sialic acid moiety on the liposomal surface and can be rejected by macrophages [32, 33]. This is very promising for design and preparation of an ideal drug carrier which can escape from the accumulation in RES at high extent.

In addition, we can conjugate a more specific monoclonal antibody or its fragment to the polysaccharide on the surface of liposomes. By this methodology, we will be able to develop more stable and targetable drug carriers for missile therapy.

Acknowledgements
The authors gratefully acknowledge the dedicated contributions of coworkers whose names appear in the references. This work has been supported by Grant-In-Aid's for Scientific Research from the Ministry of Education, Science and Culture (Nos. 56370034, 57219013, 00589012, 58211015, 58218024, 59105007 and 59212032).

REFERENCES

1. Kirby, C. and Gregoriadis, G., The effect of lipid composition of small unilamellar liposomes containing melphalan and vincristine on drug clearance after injection into mice. *Biochem. Pharmacol.* 32, 609–615 (1983).
2. Abra, R. M., Hunt, C. A. and Lau, D. T., Liposome disposition *in vivo* VI: delivery to the lung. *J. Pharm. Sci.* 73, 203–206 (1984).
3. Sunamoto, J., Iwamoto, K., Takada, M., Yuzuriha, T. and Katayama, K., Improved drug delivery to target specific organs using liposomes as coated with polysaccharides. In: *Polymers in Medicine*, Eds Chiellini, E. and Giusti, P., Plenum, New York, 1984, pp. 157–168.
4. Sunamoto, J., Iwamoto, K., Takada, M., Yuzuriha, T. and Katayama, K., Polymer coated liposomes for drug delivery to target specific organs. In: *Recent Advances in Drug Delivery Systems*, Eds Anderson, J. M. and Kim, S. W., Plenum, New York, 1984, pp. 153–162.
5. Takada, M., Yuzuriha, T., Katayama, K., Iwamoto, K. and Sunamoto, J., Increased lung uptake of liposomes coated with polysaccharides. *Biochim. Biophys. Acta* 802, 237–244 (1984).
6. Sunamoto, J., Goto, M., Iida, T., Hara, K., Saito, A. and Tomonaga, A., Unexpected tissue distribution of liposomes coated with amylopectin derivatives and successful use in the treatment of experimental Legionnaires' diseases. In: *Receptor-Mediated Targeting of Drugs*, Eds Gregoriadis, G., Poste, G., Senior, J. and Trouet, A., Plenum, New York, 1985, pp. 359–371.
7. Sunamoto, J. and Iwamoto, K., Protein-coated and polysaccharide-coated liposomes as drug carriers. *CRC Crit. Rev. Thereapeutic Drug Carrier Systems* 2, 117–136 (1986).

8. Stahl, P., Schlesinger, P. H., Sigardson, E., Rodman, J. S. and Lee, Y. C., Receptor-mediated pinocytosis of mannose glycoconjugates by macrophages: characterization and evidence for receptor recycling. *Cell* 19, 207-210 (1980).
9. Sunamoto, J., Hamasaki, H., Sato, T. and Kondo, H. *Bull. Chem. Soc., Japan*, submitted.
10. Sunamoto, J., Sato, T., Hirota, M., Fukushima, K., Hiratani, K and Hara, K., A newly developed immunoliposome—an egg phosphatidylcholine liposome coated with pullulan bearing both a cholesterol moiety and an IgMs fragment. *Biochim. Biophys. Acta* 898, 323-330 (1987).
11. Szoka, Jr. F. and Papahadjopoulos, D., Procedure for preparation of liposomes with large internal aqueous space and high capture by reverse-phase evaporation. *Proc. Natl. Acad. Sci. USA* 75, 4194-4198 (1978).
12. Fidler, I. J., Sone, S., Fogler, W, E. and Barnes, Z. A., Eradication of spontaneous metastases and activation of alveolar macrophages by intravenous injection of liposomes containing muramyl dipeptide. *Proc. Natl. Acad. Sci. USA* 78, 1680-1684 (1981).
13. Fidler, I. J. and Poste, G., Macrophages and cancer metastasis. *Adv. Exp. Med. Biol.* 155, 65-75 (1982).
14. Sone, S., Mutsuura, S., Ogawara, M. and Tsubura, E., Potentiating effect of muramyl dipeptide and its lipophilic analog encapsulated in liposomes on tumor cell killing by human monocytes. *J. Immunol.* 132, 2105-2110 (1984).
15. Phillips, N. C., Moras, M. L., Chedid, L., Lefrancier, P. and Bernard, J. M., Activation of alveolar macrophage tumoricidal activity and eradication of experimental metastases by freeze-dried liposomes containing a new lipophilic muramyl dipeptide derivatives. *Cancer Res.* 45, 128-134 (1985).
16. Kohno, S., Miyazaki, T., Yamaguchi, K., Tanaka, H., Hayashi, T., Hirota, M., Saito, A., Hara, K., Sato, T. and Sunamoto, J., Polysaccharide-coated liposome with antimicrobial agents against introcytoplasmic pathogen and fungus. *J. Bioactive Compatible Polym.* 3, 137-147 (1988).
17. Donaruma, L. G. and Ottenbrite, R. M. (Eds), *Anionic Polymeric Drugs*, Wiley-Interscience, New York (1978).
18. Sato, T., Kojima, K., Ihda, T., Sunamoto, J. and Ottenbrite, R. M., Macrophage activation by poly(maleic acid-alt-2-cyclohexyl-1,3-dioxap-5-ene) encapsulated in polysaccharide-coated liposomes. *J. Bioactive Compatible Polym.* 1, 448-460 (1986).
19. Sunamoto, J., Sato, T. and Ottenbrite, R. M., Enhanced biological activity of polymeric drugs by encapsulating in liposomes. In: *Current Topics in Polymer Science, Vol. I*, Eds Ottenbrite, R. M., Utracki, L. A. and Inoue, S., Hanser Publishers, 1987, pp. 101-115.
20. Ottenbrite, R. M., Sunamoto, J., Sato, T., Kojima, K., Sahara, K., Hara, K. and Oka, M., Improvement of immunopotentiator activity of polyanionic polymers by encapsulation into polysaccharide-coated liposome. *J. Bioactive Compatible Polym.* 3, 184-190 (1988).
21. Ellouz, F., Adam, A., Ciorbaru, R. and Lederer, E., Minimal structural requirements for adjuvant activity of bacterial peptideglycan derivatives. *Biochem. Biophys. Res. Commun.* 59, 1317-1325 (1974).
22. Lederer, E., Adam, A., Ciobaru, R., Petit, J. F. and Wietzerbin, J., Cell walls of mycobacteria and related organisms: chemistry and immunostimulant properties. *Mol. Cell Biochem.* 7, 87-104 (1975).
23. Ottenbrite, R. M., Kuus, K. and Kaplan, A. M., Characteristic biological effects of anionic polymers. In: *Polymers in Medicine*, Eds Chiellini, E. and Giusti, P., Plenum, New York, 1984, pp. 3-22.
24. Ottenbrite, R. M. and Sunamoto, J., Improved activation of macrophages by polyanionic polymers encapsulated in mannan derivative-coated liposomes. In: *Polymers in Medicine, Vol. 2*, Eds Chiellini, E. and Giusti, P., Plenum Press, New York, 1986, pp. 333-355.
25. Torchilin, V. P., Khaw, B. A., Smirov, V. N. and Haber, E., Presentation of antimyosin antibody activity after covalent coupling to liposomes. *Biochem. Biophys. Res. Commun.* 89, 1114-1119 (1979).
26. Hashimoto, Y., Sugawara, M., Masuko, T. and Hojo, H., Antitumor effect of actinomycin D entrapped in liposomes bearing subunits of tumor-specific monoclonal immunoglobulin M antibody. *Cancer Res.* 43, 5328-5334 (1983).
27. Sato, T., Sunamoto, J., Ishii, N. and Koji, T., Polysaccharide-coated immunoliposome bearing anti-CEA Fab' fragment and its internalization into CEA-producing tumor cells. *J. Bioactive Compatible Polym.* 3, 195-204 (1988).
28. Hashimoto, Y., Sugawara, M. and Endoh, H., Coating of liposomes with subunits of monoclonal IgM antibody and targeting of liposomes. *J. Immunol. Method.* 62, 155-162 (1983).

29. Fukushima, K., Hirota, M., Terasaki, P. I., Wakisaka, A., Togashi, H., Chia, D., Suyama, N., Fukushima, Y., Nudelman, E. and Hakomori, S., Characterization of sialosylated Lewis[x] as a new tumor-associated antigen. *Cancer Res.* **44**, 5279-5285 (1984).
30. Hirota, M., Fukushima, K., Hiratani, K., Oka, M., Tomonaga, A., Saitoh, A., Hara, K., Sato, T. and Sunamoto, J., Targeting therapy of human tumor cell line xenograft in nude mice using pullulan coated liposome encapsulated with adriamycin. *Jpn. J. Cancer Chemother.* **13**, 2875-2878 (1986).
31. Hirota, M., Fukushima, K., Hiratani, K., Kadota, J., Kawano, K., Oka, M., Tomonaga, A., Hara, K., Sato, T. and Sunamoto, J., Targeting cancer therapy in mice by use of newly developed immunoliposomes bearing adriamycin. *J. Liposome Res.* **1**, 15-33 (1988/89).
32. Goto, M., Arakawa, M., Sato, T. and Sunamoto, J., Specific rejection of glycophorin-reconstituted liposomes by human phagocytes. *Chem. Lett.* 1935-1938 (1987).
33. Sunamoto, J., Sakai, K., Sato, T. and Kondo, H., Molecular Recognition of polysaccharide-coated liposomes. Importance of sialic acid moiety on liposomal surface. *Chem. Lett.* *1781*-1784 (1988).

Subject index